PLUSES AND MINUSES

T0186940

Stefan Buijsman was born in 1995 and gained a master's degree in philosophy in Leiden at the age of eighteen, after which he moved from Sweden to work on a PhD. Within eighteen months (instead of the usual four years) he gained his doctorate, making him one of the youngest ever PhDs. He is currently studying the philosophy of mathematics as a post-doctoral researcher.

PLUSES AND MINUSES

How Maths Makes the World More Manageable

STEFAN BUIJSMAN

Translated from the Dutch by Andy Brown

WEIDENFELD & NICOLSON

First published in Great Britain in 2020 by Weidenfeld & Nicolson
This paperback edition published in 2022 by Weidenfeld & Nicolson
an imprint of The Orion Publishing Group Ltd
Carmelite House, 50 Victoria Embankment
London EC4Y 0DZ

An Hachette UK Company

1 3 5 7 9 10 8 6 4 2

First published in the Netherlands by De Bezige Bij as *Plussen en minnen* in 2018

The publisher gratefully acknowledges the support of the Dutch Foundation for Literature.

N **ederlands**
letterenfonds
dutch foundation
for literature

A CIP catalogue record for this book
is available from the British Library.

ISBN (Mass Market Paperback) 978 1 4746 1248 7
ISBN (ebook) 978 1 4746 1249 4

Typeset by Input Data Services Ltd, Somerset

Printed and bound in Great Britain by Clays Ltd, Elcograf S.p.A.

www.weidenfeldandnicolson.co.uk
www.orionbooks.co.uk

CONTENTS

INTRODUCTION

Let's go back in time. I'm looking at my maths teacher with a glazed expression. On a digital blackboard there is a series of formulas and a drawing of a wavy line touched by a number of straight lines. Like everyone else doing maths in their final years at secondary school, I have to learn what these formulas and drawings mean. Why? In my case because I want to study astronomy. What I don't know yet is that I'm far too impatient to be any good at that. But what if I had known? And known that in the profession I ended up in I hardly ever need to make calculations? Then I would have entered the following question into Google: what is maths good for?

The first result that Google comes up with is an article from a Dutch daily newspaper on Pythagoras' theorem and how to cut up pizzas. Wonderfully concrete, but it only illustrates a very small part of the usefulness of mathematics. Without maths I wouldn't even have been able to look for an answer to my question in Google. Or I would have ended up with an article that had nothing to do with my question. A search engine like Google can only work by using mathematics. And I don't just mean that computers work with ones and zeros: Google uses a clever piece of maths to find a relevant answer to my question. Before Google's founders,

Sergey Brin and Larry Page, devised this method in 1998, the first hit for anyone who entered, for example, 'Bill Clinton' into the search field would have been a photograph of him together with the joke of the day. If you searched for 'Yahoo' in Yahoo, the search engine itself didn't even appear in the top ten hits! Luckily, that no longer happens – and we have maths to thank for it.

And yet a lot of people today still have the same feeling I had at secondary school. To them maths is a blackboard full of formulas that you can barely understand and will never need after you've left school. No wonder it seems incomprehensible and useless to so many people. But the opposite is true: mathematics plays a very important part in our modern society and, for anyone who looks beyond the formulas, it's easier to understand than we often think. The way Google chooses information for us shows how much maths influences our everyday lives, both positively and negatively. Digital services like Google, Facebook and Twitter can reinforce existing opinions and beliefs. Today we are constantly confronted with fake news, which is very difficult to prevent. That's partly because of the way these services work. We can only learn how to combat fake news if we understand how internet services strengthen our opinions and why it's not easy to change the way they do that.

In this book, I want to show how useful mathematics is. In a certain sense, now that I understand maths better, it's aimed at my younger self. But it's also aimed at all those people who think, like I used to, that mathematical calculations are difficult and irrelevant and are happy that they don't have anything to do with them. Since I've been working as a philosopher of mathematics and think a lot about how maths works and how we learn about it, I know that it is extremely relevant to all of us, whether we have to use it in our work or not. Mathematics is about much more than formulas, which is why there are so few of them in this book. They're useful if you want to figure out a specific problem, but they often distract from the ideas behind the maths.

In this book, in order to show that mathematics is more relevant and understandable than a lot of people think, I look at various areas of maths and the ideas behind them. These areas have surprisingly many applications which everyone can easily understand, especially if we forget about the formulas. Take graph theory: search engines like Google use graph theory to order search results, but it's also used to predict how cancer patients will react to a certain treatment and to study traffic flows in large cities.

The same applies to the other areas of modern mathematics that I examine in this book: statistics and calculus. The ideas behind them are often unexpectedly simple and much more useful than you might suspect from what you learned at school. We come across statistics in the news almost every day, in the form of figures on crime, the economy, politics, and so on. Often, it's not clear what the figures actually mean and where they come from. People were already warning about the dangers of misleading statistics a century ago, with good reason. And those warnings have become even more important today.

Like graph theory, calculus is useful because it makes all kinds of applications possible without our noticing it. Since the Industrial Revolution it has been used to improve the efficiency of steam engines, allow cars to drive themselves, build skyscrapers and much more. If an area of maths has changed history, it's calculus.

But before I discuss the many modern applications of mathematics in greater detail, we have to go back to its very first beginnings. That doesn't mean searching for complex historical sums or ancient scholars, but looking at the history of mankind itself. Each of us is born with a wide range of mathematical skills, with which we can survive even without maths lessons. However, history shows us that these inborn skills fall short when people live in large groups. At a certain point, societies simply get too big to function without maths and we have to turn our attention to arithmetic and geometry. Some cultures still manage to survive without any form of maths, but they are always small societies that, for example, don't build towns and cities. The abstraction of mathematics is necessary for things like organising a community, for security, to construct houses and other buildings, to regulate the supply of food, etc. Maths makes practical problems simpler and the world around us more manageable.

The question about what maths is good for isn't only about mathematics in practice but is, in the first instance, philosophical. That's why I begin and end this book with a detour to philosophy. Philosophers of mathematics have been asking themselves for centuries what maths is and how it works, without concerning themselves too much with sums and formulas. Some of those questions still haven't been answered, but the philosophy of mathematics has come far enough for us to know what the right answers will look like.

Yet, as with most philosophical questions, you'll have to decide for yourself what you think about maths and which answers appeal to you most. And whether you're happy with how mathematics is used today. Do the advantages of Facebook, for example, outweigh the disadvantages? I leave it to you to answer that question for yourself. In the meantime, I'll try to explain what part maths plays in applications like Facebook, why they have the disadvantages that we are all now familiar with, and why those disadvantages can't be solved by simply changing the mathematical idea behind them.

CHAPTER 1

Maths all around us

Every time you use Google Maps to find the way some-where, you rely on a piece of mathematics. You open the app, enter your destination and, within a few seconds, it will automatically come up with a few possible routes. Google can only do that through clever use of maths.

Imagine that Google was crazy enough to get people who are good at reading maps to work out your route. Every time you searched for a route, they would put those people to work. Not only would that take a very long time, but it would also be very inefficient. Google's map-readers would repeatedly have to work out the same route for people like me who can't remember how long it takes to get from home to their friends' houses. Preferably, they would work out all kinds of routes in advance, just in case someone might need them one day.

But would that be any better? The chances that someone else needs exactly the same route as you aren't that great unless, for example, you live in a student house and are looking for the best route to the university. And my neigh-bours certainly won't be visiting my friends or my publisher, whom I look up in the app because I want to make sure yet again that I'll get there on time. Unless Google can predict what journeys I'm going to be making, it is regularly going to need someone to work out new routes. And, let's face it, no matter how good they are with a map, that takes a lot of time.

That's why we leave the map-reading to maths. A computer works out the best way for you to make your journey, but not in the same way that people do. The maths that a computer uses doesn't recognise streets on a satellite photo and can't calculate distances from the scale of a map. Navigation systems see the world as a collection of small circles joined together by lines. Strange as that may sound, it will be familiar to anyone who has seen a map of an underground train system, like this example of the London Underground below.

The London Underground*

For Google Maps, it's ideal if you only have to travel by

the Underground, as the map's already designed to suit its maths. The computer can now pretend it's travelling along the lines between the circles, like a virtual train. The only problem for a computer is that it can't see the whole network. If you want to plan a journey from Holborn (where the red line and the dark-blue line cross in central London) to Seven Sisters (on the light-blue line in zone 3, to the east), you can find your way quite quickly. The dark-blue line and the light-blue line both stop at Finsbury Park, just before Seven Sisters. So the best and quickest route is to go up to Finsbury Park on the dark-blue line and then take the light-blue line one stop to Seven Sisters.

A computer, on the other hand, has to go through a more convoluted procedure to figure out the shortest route. It doesn't know where Holborn and Seven Sisters are in respect of each other, so its virtual train has to travel around haphazardly until it finally arrives at the right destination. What's more, the computer needs to know how long it takes for the train to go from one little circle to another. As we are all aware, the distances shown on the lines on the underground map don't represent the true distances from one station to another. It takes slightly less time to travel from Park Royal to Alperton (on the vertical dark-blue line to the west) than from Park Royal to North Ealing (three minutes rather than four), while on the map Alperton looks much further away.

The solution to this problem is to place a number next to each line between stations in the network to indicate how long the train takes to travel along that section. The computer can then use those numbers to find the best route. One of the simplest navigation systems will look at all the different routes from Holborn, starting with the shortest possible route, then the next shortest, and so on.

So in this example, the computer starts at Holborn and looks for the nearest station. Both Chancery Lane and Russell Square are only one minute away, so either of those could be the first option. Does the computer then continue on to King's Cross from Russell Square, or to St Paul's from Chancery Lane? No, it makes a second attempt to travel in the direction of Covent Garden, as that is shorter than the journey to King's Cross or St Paul's. And then it looks at Oxford Street, which is three minutes away from Holborn. Only after looking at all these options does the computer move on to the second station from Holborn in the direction of Seven Sisters.

In this way, it takes the computer quite a while to finally end up in Seven Sisters, a twenty-two-minute journey passing seven stations. Before it gets that far, it's already been to Brixton, only nineteen minutes away at the bottom centre of the map, and to Edgware, even shorter at seventeen minutes, at the top centre. But it gets there in the end and, once it has found Seven Sisters, the computer is sure that the first route it has calculated is the shortest. It all sounds far from efficient, and the human sense of direction and ability to oversee the whole picture seems much simpler. Yet a computer is still quicker than we are, simply because it can calculate far more steps per second.

Google Maps works in much the same way. The little circles aren't stations but road intersections, like a roundabout or a motorway junction. For the maths, whether a line is a motorway or a side street makes a difference. Just like the Underground map, it all comes back in the information you are given on journey times, based on the numbers that Google Maps places next to every line. A section of motorway and a side street may be the same length, but you have to drive much more slowly on the side street, so in the system the number next to the street is much higher than for the motorway section. The numbers also help to adapt journey times to take account of congestion. All Google has to do is raise the number alongside the congested road from, say, ten minutes to twenty to account for an expected delay of ten minutes. If you then recalculate your route, the delay will be automatically included in the new journey time. You might be redirected along side roads to avoid the congestion, because a route that was at first longer is now quicker.

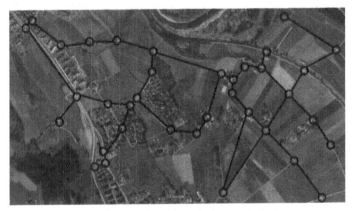

A road network as Google Maps sees it*

This method works very well over short distances, but the maths becomes unmanageable if you want to travel longer distances. If you want to drive from New York to Chicago, Google first runs through all possible routes from New York that take less than twelve hours (the time it takes to make the journey). Computers can calculate very quickly, but even a modern computer can't make that many calculations in a

* Image © Elias Wirth

short time. So Google Maps uses a number of mathematical tricks to reduce the number of calculations. We don't know exactly what they are, as they keep their methods secret, but we'll look at them in more detail in Chapter 7.

As we have seen, the routes recommended by navigation systems are shot through with mathematics. But that maths is not necessarily cleverer than we are. The computer's desperate search for the final destination is often anything but efficient. The maths doesn't make the problem any simpler, because in the end the computer has to do more work than we do; but it does make life easier for us. It can work out the best route more quickly because it can do such an astounding number of calculations per second.

Recommendations from Netflix

After you've got Google to work out what route you should take, you leaf through the new films and series on Netflix. Next to each one, in green, is a percentage that tells you how well it fits in with what you normally watch. Sometimes it's completely wrong and a film you are supposed to think is wonderful turns out to be a great disappointment. But if, for a change, you don't ignore the percentages, they should reflect your tastes quite accurately. The recommendations are generated fully automatically and will change the more you watch other kinds of programmes. In other words, there's a computer program somewhere that knows, without having any understanding at all of films and series, what suits your tastes and what doesn't.

Netflix does this, of course, on the basis of information it has about its users. Massive numbers of people watch films and series on Netflix and the company keeps a record of their viewing habits. In very simple terms, that means that

Netflix knows what kind of movies and series we all watch, whether they are documentaries about route-planning algorithms, horror films or something else. Netflix also puts all its films and series into categories. Then it uses two sets of data to make its recommendations. If you watch a lot of horror movies, you will probably want to watch one you haven't seen before. Sounds simple enough.

The difficulty lies in what else Netflix does. It gives all films and series which don't fall into a certain category – in this case, horror – a score in the form of a percentage. The percentage shows how well the film fits in with what you normally watch. In other words, Netflix also decides how similar an adventure film is to a series of horror movies. If a lot of scary things happen in it, it will fit in better with your normal viewing behaviour than one which is less horrific. These are the kinds of details your friends often tell you if you ask them to recommend something to watch. Netflix can give you that information too, though its recommendations are by no means as accurate as those of a real movie buff.

What makes it more complicated is that you might only watch certain kinds of horror films. If you don't like ones with a lot of blood, very bloody movies will fit your profile much less closely than a somewhat scarier-than-average adventure film. Simply looking at the general category doesn't always produce the best recommendations, since what really matters is the content of the movie. As computers have no understanding of content, perhaps Netflix should simply hire lots of staff who do. But, as that's not feasible with millions of viewers, it has to rely on its computers and algorithms to make their recommendations. That's possible, but it does require a trick.

The idea is actually very simple: a recommendation is

good if it's similar to what you like to watch. All around the world, people watch programmes on Netflix that they like because they are similar to series and films they have seen before. For Netflix's computers, two films are similar if a lot of people who have seen one go on to watch the other. If thousands of people watch *Iron Man 2* after having watched *Iron Man*, they must be similar, and *Iron Man 2* is a good recommendation for people who have seen *Iron Man*. The more people use Netflix, the more accurate the predictions will be. The computer program suggests films and series that a lot of other people have seen and which are roughly similar to what you have watched yourself.

This solution comes with a problem. Netflix has millions of users, each of whom has watched a large number of movies and series. The trick Netflix uses to make recommendations is a simple mathematical calculation: it looks at how many people with the same viewing history have also seen the programme it wants to recommend. The problem is in the calculation. I explain it here in a simplified form, partly because the details are not public. Netflix also has to take account of people who have seen almost the same programmes, but not quite. And what about people who like not only horror films but also documentaries? There are then far fewer people who have seen the same programmes, and that makes the recommendation less reliable. The simple idea proves to be a lot more complicated in practice.

That's why it helps to show all the films and series on offer on a map like the one of the Underground network we looked at earlier. Every film or series is a circle, like a station in the world of Netflix. You can travel from station to station by clicking on two different films or series on the Netflix website.

On this map, too, numbers have to be added to enable the calculation to be made. The numbers represent not travelling times, but how many people have seen both of the programmes connected by a line. You can see a very simple example below, with three movies and fictitious numbers showing how many people have seen them in different combinations:

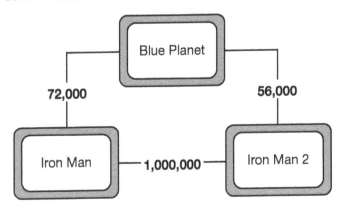

Netflix: fictitious viewing figures for three films

The question is: what percentage should each film get, as an indication of how well it fits in with your viewing behaviour? Let's say you've only seen *Iron Man* on Netflix. The computer has to predict how much you would like *Iron Man 2* and *Blue Planet*. According to the figure, *Iron Man 2* should get a very high percentage. After all, you are much more likely to enjoy a film that a lot of people who share your taste in films have also seen. *Blue Planet*, on the other hand, should get a low score as few people who have seen it have also watched *Iron Man*. What's more, only a few people who have seen *Iron Man 2* (which the computer thinks would suit your taste) have also watched *Blue Planet*

– another reason to give *Blue Planet* a low percentage.

Ultimately, a computer uses its own predictions – for example, of how much you will enjoy *Iron Man 2* – to improve its predictions for other films and series. That's not too difficult to keep track of with only three films. But try doing it with thousands of films and series. In theory it is possible; with enough time and space, you can also work out for yourself every route you want to travel. But thanks to maths, and especially the graphs that we will look at in greater detail in Chapter 7, it's not only possible in theory, but is feasible in practice if you have a computer that is powerful enough. The mathematical version of this puzzle makes it possible for Netflix to predict fully automatically whether you will enjoy a particular movie or series, without the need for an army of movie buffs.

Maths is everywhere

We come across maths every day, in all kinds of places. Not literally, of course; even I don't have to do any mathematical calculations on an average day, though it is my job to think about maths. And yet, behind the scenes, maths plays a major role in our lives. Without it there would be no Google Maps to show you the way. Netflix could suggest a few films and series you might like at random, but they would be far less targeted to your specific tastes. Google's search engine would hardly work at all. In short, services that we use every day are only possible because they are using maths in the background.

Netflix, search engines and route-planners are examples of services that depend on the same branch of mathematics: graph theory. But this is not the only area of maths that is important on a daily basis. Many of the news articles that your telephone alerts you to contain statistics. Election polls, for example, claim to show the voting preferences of a whole country in a series of numbers. But how useful are they? They can be way off the mark, as we saw in the 2016 American presidential elections. According to the polls Hillary Clinton was going to win, with some experts claiming to be almost 100 per cent certain of the outcome. Figures can easily be misleading, even if that is not the intention. Statistics conceal all kinds of things. For anyone who doesn't understand what can go wrong, an important-sounding statistic is as good as useless. It's all very well that polls tell you something, but how can you trust them if they are so often completely wrong?

You look up from your telephone for a moment to order an espresso. It is made using a large stainless-steel coffee machine that heats the water to exactly the right temperature. If it's a luxury model, it doesn't happen just like that. The machine monitors how quickly the water heats up and, on the basis of this information, decides whether the water needs to be heated a little more or allowed to cool off, and so on. It keeps doing that until the water reaches the perfect temperature and it can make the coffee. You don't see any of this happening but, right in front of your nose, the formulas your maths teacher used to talk about are being used to make you a cup of coffee.

While you drink your espresso, you read the political news. The government has changed some of its policy plans. You're not sure whether that was such a good idea, so you take a look at the predictions for the new plans. As always, economic research institutes have analysed them in detail. Whether something is a good or bad idea can depend on so many factors that you can hardly keep up with it all. But a single calculation combines all those factors together in one piece of information that is important to you: whether you will ultimately have more money in your pocket. And that also depends on a lot of mathematical calculations.

Looking at it in this way, we see that maths has a great influence on our lives. Without actually doing any ourselves, we are dependent on a whole lot of calculations. The information we use to make choices is the end result of mathematical work done by others. Even the information you ultimately get to see depends on a calculation made somewhere on a computer used by Google, Facebook or another website that filters data. The technology all around us uses maths more and more. Not only that luxury coffee machine at the café round the corner, but also the automatic pilot in the aircraft that takes you to your holiday destination and the computer you use day in, day out to do your work – they all depend on mathematics. With maths being used all around us, it's increasingly important to understand it better and how it affects our lives.

That is largely what this book is about: how useful it is to know a little about maths. But what is maths and how does it work? That is above all a philosophical question which goes back to Plato and Socrates. They asked themselves what mathematics is about and how we can learn about it. What's more, if you consider it a little longer, it's strange that maths is both so widely applicable and yet so abstract. How can it possibly be so useful? To answer that question we need to turn to philosophy.

CHAPTER 2

Separate worlds

A group of prisoners are chained to a wall. Their heads are fixed so that they can only look ahead, at another windowless wall. They've been chained to the same wall all their lives and, as far as they are concerned, the only reality is the shadows on the wall in front of them. If they could get close enough, they would be able to touch them. None of them knows of a life beyond their prison. Their world consists entirely of shadows.

This is how Plato's Allegory of the Cave begins. He compares us to the prisoners. The things we see around us are only shadows, cast by something that we cannot see directly. The table you are sitting at, for example, really exists, of course. Yet Plato regards it as one of the shadows on the wall. He is not interested in that specific table. What concerns him is the abstraction that links all tables, what makes the thing in front of your face a table and not something else. You can't simply see that abstraction. You have to find out what it is, what is casting the shadow on the wall. You do that by looking at all the different kinds of tables around you.

According to Plato, this is also how mathematics works. This is his answer to the question: what is maths about? For Plato, numbers are one of the things that cast shadows. We can't simply look at them. A number is not something you can hold on to or bump into as you are walking and looking at your telephone. I can of course write down a number, like '2' for example, but just as the word 'sun' is not a star, the number '2' is not the same as the number I am trying to talk about. To stick with Plato's example, the world around us is nothing more than shadows, while the numbers are always somewhere behind us.

We can see maths in this way. If we talk about numbers and say, for example, that $1 + 1 = 2$, we are talking about things that really exist. But they don't exist in the same way that the table in front of your face does. Plato saw them as 'more real' because he thought abstract knowledge was more worthwhile than knowledge about specific things. That is why he reduced the objects that people normally see around them to shadows and believed that numbers and other 'real' things float around in an alternative universe. I think that's going a bit far, but his idea that numbers exist has been so influential that we still call people who agree with him 'Platonists'.

So is that mathematics? It may sound logical to think of it in that way. Your maths teacher tells you what the world of mathematics is like. It's a real world, you just can't see it. Mathematicians study that world, just as physicists study the world we can see. Then mathematics seems far away from our ordinary lives. No wonder so many people have a problem with maths: you first have to work out how to find that other world before you can learn anything about it.

How do you learn something about a world that you can't see, feel, smell or sense in any way at all? According to Plato and the Platonists, mathematics is separate from everything in our daily lives. Although not entirely: Plato uses the example of a slave in his friend's household to show

how we can come into contact with mathematics. The slave, who has had no education, is asked to draw a square twice the size of one that has already been drawn in the sand without using measurements. That's quite a tough problem. If you enlarge the square by making all four sides twice as long, you end up with a square four times as large as the original. To solve the problem without a ruler, you need a clever solution.

In Plato's example, people around the slave ask him all kinds of questions. The questions are cleverly chosen so that the slave works out that the trick is to start by drawing a diagonal line through the original square. In the figure below, the original square is shown in light grey. You draw three more squares the same size round it, forming a large square four times bigger than the original. You then use the dotted lines shown in the diagram to cut the large square down by half. That leaves you with a square exactly twice the size of the original.

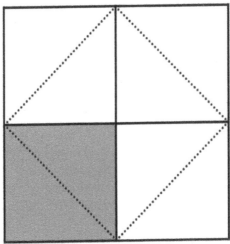

How to make a square twice the size of the original

In this example, Plato only helps the slave by asking questions. The slave gradually finds out 'for himself' how to double the size of the square. Plato, who wants to show us how maths works, says that the slave actually knew the answer already; the philosopher only helped him to remember it. This is because, Plato claims, in a previous life we knew everything there is to know about maths. That knowledge is still there, deep down in our subconscious. To learn something about maths all we have to do, according to Plato, is remember what we already knew.

Does that sound a little far-fetched? I hope so, because Plato's solution is pure nonsense. He cheats with the questions in his example. He draws the solution first and then asks a series of yes/no questions, to show that the trick with diagonals works. The slave 'teaches himself', but only because it is explained to him one step at a time, in the form of questions. And then to claim that he remembers all these things from a former life: that must have been a very special life, to say the least.

So if Plato's solution is nonsense, how do we learn about the world of mathematics? The truth is, we don't really know. Present-day Platonists – the people who believe that maths is about real numbers – answer this question in a different way. Whether any of them are right is a different matter. They do, of course, think that we are able to learn mathematics: after all, we have just learned how to make a square twice as large as the one we started with. Even the most hopeless of pupils knows at least something about numbers. Platonists just haven't figured out how that works, how we can learn something about that inaccessible, abstract world of maths.

But why should we think that the world of mathematics is so inaccessible and abstract? Plato thought that because a lot of mathematicians in his time said that was the case, and their present-day counterparts still do. But should we believe them? There's a whole group of modern philosophers who say that it would be better not to believe them. Forget the prisoners in the cave and think of Sherlock Holmes.

Maths as one big story

Sherlock Holmes lived in London, at 221b Baker Street. You can visit the house. But of course he didn't really live there. Sherlock Holmes is a fictitious detective. A lot of stories, films and television series have been made about him. That's why we don't immediately think, 'What nonsense, how can Sherlock Holmes have lived in Baker Street?' In the stories he does live there, but no one of that name has ever lived at that address in the real London. We can think of mathematics in the same way.

Maths tells a story, about numbers, figures and all kinds of other things. It's about a world like the one Plato described. Nothing ever changes and everything fits together in a perfectly logical way. Nominalists, however, say that is just as much fiction as the stories about Sherlock Holmes. The world that maths talks about doesn't exist. Mathematicians talk about things like numbers and triangles, but those things are not actually there. The only thing that is real is what you see around you; there is no separate world with figures floating around.

In other words, Plato claims that we discover things about mathematics that are already there. But perhaps there's nothing to discover and we have dreamed up everything about maths ourselves. That produces the most absurd consequences: if the things that maths talks about don't exist, then nothing is true about all those numbers, triangles, etc. either. We say that 3 is a prime number and $1 + 1 = 2$, but that's not true: $1 + 1 = 2$ isn't true because numbers don't exist. Just as it's wrong to say that Sherlock Holmes lived in London because he never existed.

So why can't you just tell your maths teacher that everything in mathematics is nonsense? Because something about maths is most definitely true, even for nominalists. My sentences about Sherlock Holmes, for example, are true in a certain sense. They are consistent with the story by Holmes's creator, Sir Arthur Conan Doyle. If I were to claim that Holmes lived in Alaska, you could prove with passages from his books that I was mistaken. You can always test whether statements about Sherlock Holmes are in line with what's in the books. And the same applies in maths: the claim that $1 + 1 = 3$ is untrue in terms of the story of mathematics.

And yet, I repeat that we don't know exactly how mathematics works. In other words, we don't know whether it makes discoveries about an abstract world that is difficult for us to access, or whether we have simply made it all up. That is because neither the Platonists nor the nominalists have succeeded in explaining how we learn about maths.

For Plato, it was difficult to say how we can access the abstract world of numbers. That's not so much of a problem if that world doesn't exist. It is, for example, blindingly obvious how we learn about Sherlock Holmes: by reading a book. If we can remember what we have read, we have learned something about the detective. But with maths, that's a little more difficult. The stories in maths are special because we usually believe that they describe the world around us, while no one takes the books about Sherlock

Holmes literally. That makes it difficult to explain how maths works. How can you be sure that people who say things that are literally untrue nevertheless manage to build up a science that is valuable and rigorous? Nominalists still have no good answer to that question.

So much for philosophy. Don't worry if you can't follow all the details that philosophers make so complicated. What I mainly want to show you is that we can look at maths in two ways. No matter how differently the Platonists and nominalists think about mathematics, they both try to explain how it works, or tell us what happens when we are 'doing' maths. Platonists say that we discover all kinds of knowledge about a world full of abstract things. Nominalists say that that world doesn't exist and that we are just making everything up. As long as that difference is clear and you're willing to believe that we still don't know which of the two is true, you know enough.

The value of beauty

If all the squabbling between philosophers makes one thing clear, it is that mathematics is about something very abstract. No wonder, then, that at secondary school it's often unclear what it's good for. Maths seems to have nothing to do with the world around us. Not if you see maths as a world in itself that doesn't come into contact with the physical world at all. And not if you think of maths as a story, as what does that story have to do with our real world? After all, we don't read the stories of Sherlock Holmes to learn something about historical London, so why should we do that with the stories about maths? How can mathematics, which has nothing to do with the physical world, nevertheless be used to understand that world?

Maths can actually be very useful in helping us understand the world around us. In Chapter 1 we looked at a number of examples where mathematics plays a major role in making problems simpler. It helps precisely because it is abstract – and it doesn't only do that in everyday life. Scientists, too, have been using maths for centuries to make new discoveries, which are extra-surprising because they show us that maths is even more useful than we might think from the examples in the previous chapter.

To start with, there is the story of Isaac Newton (1643–1727). As a young man, Newton was sitting under an apple tree staring at the countryside around him. Suddenly an apple fell on his head and Newton thought, 'That's it! It's gravity!' At least that's how the story goes. Apple or not, Newton's ideas on gravity were groundbreaking. For the first time in history, someone had realised that we can explain why things fall downwards on Earth in the same way as we explain the movements of stars and planets. The rest is history. We all know that Newton's idea was brilliant and not just a crazy theory about two completely unrelated things.

But Newton's contemporaries thought just that. Newton described gravity as a force that acted from a distance and made things attracted to each other in an almost magical way. In his time, people believed that everything happened because things came into contact with each other. That's not such a strange idea: how can things influence each other if they don't interact in some way? How can the Earth 'know' that the Sun is there and is pulling the Earth towards it if the two bodies are not in contact with each other in any way at all? Thanks to Einstein, we now have a reasonably good answer to that question, but when Newton proposed the theory of gravity there were no good answers, just an impressive piece of mathematics. The question was whether it was also right.

We know that Newton was largely right, thanks to the predictions of his theory. Now that we can measure everything more accurately, we can see that they match reality very well. In Newton's own time it was far from obvious that his theory was the best one. What scientists observed could deviate by as much as 4 per cent from Newton's predictions, and yet he felt that a theory that applied to both the Earth and all other celestial bodies was better. Such a theory is 'neater', not only simpler in terms of physics, but the maths is also less complicated.

The surprise lies in what happened next. Physicists continued to test Newton's theory. And with the instruments we now have at our disposal, which are of course much more accurate than anything Newton could use, we know that his theory is never wrong by more than 0.0001 per cent. Although he wasn't aware of it, his preference for a single theory based on a simple piece of maths was a great success. Despite the fact that Newton had no way of making precise measurements when he came up with his theory under that apple tree in the English countryside, his mathematical predictions have proved enormously accurate.

Sceptical readers may claim this was all coincidence. Newton was just lucky, unlike all those others whose names we have forgotten. Perhaps it was coincidence, but there are too many stories like this to ignore. Copernicus came up with the model of the solar system that we still use today: with the Sun at the centre and the Earth in orbit round it. His model also used simpler, more elegant maths than the increasingly complicated models with the Earth at the centre and the Sun moving round it.

Copernicus's predictions were actually less reliable than those more complex theories. That's because he thought the Earth's orbit was a circle, while in reality it is an ellipse. But, in the end, his simpler theory – the one that mathematicians prefer – proved the better one.

More remarkable is a discovery by Paul Dirac in the early twentieth century. Dirac was working on quantum mechanics. His aim, like Newton's, was to explain different aspects of physics in the same way. And he, too, did that using a mathematical model that produced the right results for what was known at the time.

But Dirac had a problem. Although his model did the job, it also made strange, extra predictions. One thing Dirac was interested in was the electron, the small particle that revolves round the nucleus of an atom. Physicists at the time already knew quite a bit about electrons and Dirac's formula described their behaviour very accurately. However, according to the formula, there should be another particle, the exact opposite of an electron. No one had ever seen such a particle and there was no reason to believe it existed.

We know now that Dirac's mathematical model had produced a completely new prediction. But back then, at the beginning of the twentieth century, it took a while for Dirac and other physicists to understand what was going on. Initially, Dirac proposed that the mysterious opposing particle was a proton. Protons had already been discovered and have a positive electric charge, while electrons have a negative charge. But that wasn't possible: protons are much heavier than electrons, so can't be their exact opposite. Dirac saw no other solution than the presence of an extra particle: a positron or anti-electron.

This is something we haven't yet come across in the

examples we've looked at so far in this book. Here, maths hasn't only made a problem simpler or made better predictions than expected; it has also predicted the existence of something we've never seen before. Scientists started looking for that new particle purely because Dirac's maths seemed too good to be wrong.

And they struck gold. Not long after Dirac's prediction, Carl David Anderson proved the existence of positrons. In 1936, only four years after his discovery, he was awarded the Nobel Prize. Positrons are not only the opposite of electrons; they are the first anti-matter particles ever to be discovered. And their discovery was made possible by mathematics.

There are more discoveries like this in physics, where something strange in maths eventually proves to be right. Around 1823, Augustin Fresnel was interested in the behaviour of light. He, too, was a physicist who devised a simple and compact mathematical formula to explain something in the world around us: what happens to light when it is reflected, for example by a mirror.

A mirror is an easy example, because the light is reflected very accurately and predictably. If you stand directly in front of a mirror, the light comes straight back at you and you see yourself. But if you look at the mirror from an angle, you don't see yourself but something that is exactly the same distance away on the other side.

Fresnel was more ambitious. He wanted to know what happens, for example, when light from water comes into contact with air. Or when light from the air strikes transparent glass. That sounds difficult, but the formula that Fresnel came up with is hardly more complicated than that for mirrors. He only needed to add one symbol. It was another simple and compact piece of mathematics but, as with Dirac's model, it could produce strange results.

Fresnel's formula predicted that light can sometimes be bent at an impossible angle. His mathematical model used complex numbers, 'extra' numbers that don't refer to things that can be counted in an ordinary way. Back then, they were seen as necessary to make the calculations easier, but weren't to be taken too seriously. When Fresnel's calculations came up with one of these numbers he panicked: his compact model was predicting something completely impossible!

As he didn't want to abandon his model, Fresnel decided to assume that the strange result was correct. And he was right: in those cases where the model threw up strange results, something very exceptional happened to the light which was completely in line with the calculations – and that meant that they weren't so strange after all. Even when light goes from water into the air, it is perfectly reflected, just as though the water's surface is a mirror. What Fresnel had discovered with his mathematical model was something that physicists had not really thought about till that moment. But it was something we all recognise. Look at how the turtle is reflected in the water's surface in the picture. That strange result produced by Fresnel's model, the complex number that no one wanted, describes that reflection. The compact mathematical formula is once again right; the strange result shows us something we failed to see before.

Reflection of a turtle on the water's surface

These examples illustrate how maths can be useful in so many ways. It makes problems easier, and enables physicists to discover new things. All because they prefer compact mathematical models, and accept occasionally strange results as part of the package. Even if there is no evidence at all that the maths is correct, they stick to their formulas – and often with good reason, as proves to be the case time after time.

Of course, it doesn't always work out that well. There are plenty of theories that prove to be wrong, compact or not. But what is surprising is just how often it does work out: that a mathematical model that scientists like because it is so compact also helps us to understand the world better. For anyone who thinks about maths, in whatever way, that is a puzzle that has to be solved. Maths is clearly not useless, but how is it possible that we can apply it so successfully?

Let's go back to Plato's abstract world. That's far from our everyday lives, a world of numbers that has nothing to do with us or the world that physicists try to describe. The maths that makes predictions about the world around us doesn't come from that physical world, so the predictions, too, seem to come from nowhere. How can the world of mathematics know anything about the real world?

Looking at Sherlock Holmes stories doesn't help us answer these questions either. Stories may not be as remote from the world as Plato's numbers, but they are and always will be fabrications. The mathematics that Newton used was devised before we knew that it could help us so effectively to understand gravity. Dirac made his formulas before anyone knew that positrons existed. In the same way, it would be strange to find out that things about London in the time of Sherlock Holmes were true, when Conan Doyle had only made them up and put them into his stories because they fitted in well with the plot.

That is why it is so fascinating that mathematics actually works. I can give an enormous number of examples that show how useful maths is. In the following chapters I am going to do just that, focusing on how it directly influences our daily lives. In the final chapter, I will return to a question that has fascinated me since I started out as a philosopher: how is it possible that maths works?

But that isn't the most important question I address in this book. First, we have to establish that maths is useful, and that it is important to know a little about it yourself. After all, why should it matter to you, knowing that maths is useful, if you don't need to use it yourself? Can't you just lead a happy life without having anything to do with maths at all?

CHAPTER 3

A life without numbers

Under a cloudless blue sky a man is sailing down the Maici, a river in the heart of the Amazon rainforest. On the banks of the river lives a small tribe that has almost no contact with the outside world. The man visits them every year, in the hope of coming back with as many Brazil nuts and as much rubber and other natural products as possible. His boat is filled with the usual goods he brings to trade: whisky, tobacco and more whisky.

Trading with the Pirahã is quite a challenge. They've been doing it for 200 years and still speak only a few words of Portuguese. Luckily that's enough for him to get what he's come for: valuable nuts and rubber at a price to make other traders green with envy. That price can, however, vary widely: sometimes the Pirahã will exchange a full bucket of nuts for a cigarette while, at others, they'll demand a whole packet of tobacco for just a handful. Other than that, it's all very simple: the Pirahã pick out goods on his boat until the trader starts to protest.

The tribe has a completely different way of looking at the whole business. While the Brazilian traders have no idea what price they'll get for their tobacco or whisky, the Pirahã don't see that a problem. They have no concept of numbers. They don't stick to a fixed price simply because they don't know how to. And they don't see any reason to do so, though they do have a clear image of the various traders in their minds. They all know who's honest and who will always try to give them less than their goods are worth.

All this was discovered by Daniel Everett, an American researcher who has lived among the Pirahã for many years. He is one of the few people from outside the tribe who speaks their language. Everett discovered that the Pirahã have no words for numbers. They sometimes refer to rough amounts but don't even have a word for 'one' (they also have no word for 'red' or a way to express the perfect tense). That makes the Pirahã one of the few cultures that don't use maths at all. Their language (like a handful of others) has no words for lines, angles or other geometric concepts. Since mathematics has only existed for 5,000 years, this extraordinary society offers us a unique glimpse into our own past.

That makes the cultural differences between us and the Pirahã enormous. They are not concerned with keeping track of how much things are worth, what time it is, or whether they have enough money to get through to the end of the month. They don't have money; they trade by barter. All this is possible because they live in very small groups. Everyone knows each other, and only the living are important. They don't keep track of their family trees and when one of them dies, once all the people who knew them have died too, they are forgotten. The lives of the Pirahã focus completely on the here and now.

There's not much call for mathematics in a culture like that. At the insistence of the Pirahã themselves, Everett tried to teach them maths for a while, but failed spectacularly. Every day for eight months he gave them lessons about numbers and geometric shapes and asked them to draw a straight line or to put the numbers 1 to 5 in the right order.

And yet, in all that time, he couldn't teach them anything at all about maths.

Are they just not able to learn it? They probably can, but the Pirahã don't seem interested in knowledge from outside. They don't believe that questions can have right or wrong answers. When Everett suggested that there might be wrong answers to mathematical questions, they drew a few signs on paper or named a few random numbers. Sometimes they ignored the maths completely and talked about something that had happened that day. Even being asked to draw a straight line twice in a row was too much for them.

It all sounds like my own maths lessons, except that the Pirahã were there of their own free will. Though not so much for the maths: Everett always made popcorn and it was a good opportunity for everyone to get together and catch up. Maybe it wasn't so different from my secondary school after all.

On an island, far away from the Pirahã

There are only a few cultures left in the world that don't use maths. The Pirahã are an extreme case because they don't even have words for numbers, but in Papua New Guinea there are a few communities that do have such words but hardly use them. They, too, survive without using maths at all.

The Loboda live on Normanby, a small island to the east of the larger islands of Papua New Guinea. They count using parts of the body: their word for '6', for example, is, literally translated, 'a hand and a finger from the other hand'. But it's not very useful because in situations where we would use numbers, they see no reason to.

Take money, for instance. We use money to buy things.

Everything has a price, expressed as a number. The Loboda do have money: coins and notes, which they can exchange for euros or pounds. But they can't give someone money as a present at one of the many parties they organise. When they receive a gift, they have to give back exactly the same gift later. If a neighbour gives them a basket of yams at a party, they have to give him a basket of yams of exactly the same size at a later party. It can't be money or something else of the same value; it has to be just as many yams as he gave them.

To us, 'just as many' suggests the exact same number of yams. But the Loboda never count the number of yams in a basket; they only make a rough estimate. They look to see if the basket is full or only half full. If it isn't completely full, you can give a little more or less back without it making a difference.

The Loboda don't use numbers in other situations, either. We soon resort to numbers when we talk about age or length or time: how many years old someone is, how many centimetres or inches long something is, or how many minutes ago something happened. The Loboda talk about these things too, of course; but they will describe how long something is, for example, by comparing it to something familiar. A chain can be as long as a forearm. That sounds similar to our foot, but for the Loboda it's not a unit of measurement. It's nonsense for them to say that something is two feet or two forearms long. If something is longer than a forearm, they'll simply compare it to something else.

To describe someone's age, the Loboda say they are as old as a baby, a child, and so on. And they describe time in the same way: something may take as long as a journey from the village to the next island, for example. They survive perfectly well without numbers.

The Yupno, another tribe from Papua New Guinea, couldn't agree more. Their villages lie some 2,000 metres up in the mountains in the province of Madang. Like the Loboda, they count by using parts of the body. Although they don't always do it the same way, generally speaking it goes as shown in the drawing below. To express a number, you say the word for the corresponding part of the body or point to it – easy enough if you're a man but, as the drawing shows, at some point women will have problems with this system of counting.

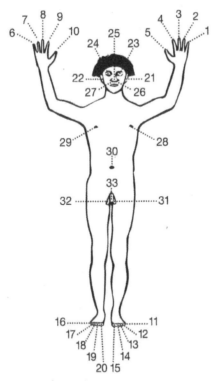

The Yupno counting system

The Yupno sometimes also use sticks to count, laying them down one at a time. As they don't live in such great isolation, most of the younger members of the tribe have had a Western education and count using Tok Pisin, a language similar to English.

The Yupno therefore have three ways of counting, but they don't use them regularly. They have given everything a fixed value, which is not expressed in a certain number of coins. Instead, they present all their goods in piles that are worth exactly one ten-*toea* coin, one of the smaller denominations in their currency. Piles of tobacco are smaller than piles of food and you can't buy only one banana – you have to buy as many as your coin is worth. This saves them having to mess around with small change and there's hardly any need for them to count.

There is, however, one important exception: the dowry. Among the Yupno, dowries consist mainly of pigs and money, which they count in two ways. Some of the men count out loud, using parts of the body, and others lay out sticks. That is to avoid confusion, since not everyone counts in the same way. If, for example, they go directly from the hands to the ears, the right ear means 12, rather than 22 as shown in the drawing. When that happens, the sticks are a good way of keeping track.

Since the Yupno take such great pains to count their dowries precisely, researchers thought they could use dowries to teach them maths. They asked one of the older members of the tribe: 'You need nineteen pigs for a dowry and you already have eight. How many do you still need?' His answer was surprising: 'Friend, I'm not rich enough to buy a new wife. Where would I find eight pigs? Besides, I'm an old man and have no more strength.'

No measurements required!

All in all, these tribes can survive just fine without using numbers. But don't they need numbers to measure things? Don't they need at least some understanding of numbers, lengths and distances to build things or find their way? Apparently not. The Pirahã, the Loboda, the Yupno and many other cultures can do all this without using maths.

Several of the tribes in Papua New Guinea build canoes. As the country consists mainly of islands, they don't have a lot of choice. It is – or rather, it used to be – the only way to travel from one island to another. The Yupno don't need canoes since they live high up in the mountains, but the coastal tribes need sturdy boats that will not suddenly sink while they are at sea. To build them, they have no blueprints with standard measurements and requirements for the thickness of the tree trunks. They use their experience, comparing their new canoes to ones they've built before.

They do back up that experience with a simple form of measurement: not a measuring tape or a yardstick but a forearm or, on the Kiriwina Islands, thumbs and the palms of their hands. That makes the people of the Kiriwina a little more accurate in their measurements, and that's just as well since the islands are small and they spend a lot of time at sea. So they measure the size of their canoes very carefully, though they never change the basic shape.

More important than the size and shape of a canoe is the thickness of the wood. If it's too thin it can easily be damaged, and if it's too thick the canoe will be able to carry less weight. Not that the tribes in Papua New Guinea use very accurate methods to measure the thickness of the wood. Some use their legs, while others have discovered that you can hear whether the wood is thick enough – and the canoe is therefore safe – by giving it a sharp tap. Often they don't know how much weight a canoe can take until it's in the water.

All kinds of things have to be built on land, too – a bridge across water or a ravine, for example. You obviously can't test a bridge beforehand or tell from looking at it whether it's safe enough to walk across. How these people have discovered when a bridge is strong enough remains a mystery. They've been doing it for so long that no one remembers how their ancestors went about it.

The Kewabi, who live in the middle of the main island, build bridges using no accurate measurements at all. They estimate the distance from one side to the other and then search for tree trunks that look long enough. The same goes for the poles that carry the weight of the trunks. Just like the Golden Gate Bridge in San Francisco, which is also held up by pillars and cables, the poles have to stick out high above the bridge. Then they need ropes that are long and thick enough, and so on. The Kewabi have little trouble building their bridges with nothing more than good powers of estimation and a lot of experience.

Many tribes in Papua New Guinea also build their houses using a combination of estimation and experience, though how they do it varies much more widely: some, for example, build square houses, while others build only round ones.

The Kâte, who live in Finschhafen in the east of Papua New Guinea, build rectangular houses. They start by making two ropes – one for the length of the house and the other for the width. When they are gathering building materials, they use the ropes to see if they have enough. That saves a lot of work – they don't want to cut down more trees than they need.

Not all tribes are such accurate builders. A tribe in the

province of Madang build their houses without using ropes or any other measuring aids. They have a standard procedure: they set out nine or twelve poles in a rectangle, all roughly the same distance apart, for the foundations. Then they build the house on top of the poles, using only their powers of estimation.

The people in the village of Kaveve also build their houses on poles, but they are round. The entrance is a round hole in the edge of the circular floor, leaving room in the centre for a fireplace. They use ropes to work out how large the two circles should be. The entrance hole should be as small as possible, to prevent draughts. So they measure the fattest person in the village and, if he or she just fits through the hole, it's big enough. The people of Kaveve therefore do use measurements, but only to a limited extent. No one works out how much wood is needed or how large the surface area of the house will be. They still gather building materials and do the actual building by intuition. The ropes tell them how big everything has to be, but that's all. You can build houses, bridges or canoes without using maths.

Working with small quantities

All kinds of cultures, therefore, hardly use maths. Even if they can, even if they have a system of numbers, they don't need to use them. They can estimate lengths and quantities quite accurately. That saves a lot of time and works just as well. But how is it possible? What makes us able to trade, provide enough food, build bridges and so on without using maths? In recent decades, science has found the answer to that question. We use certain parts of our brains to work with quantities. That's why we can estimate lengths or recognise a square, even if we have never learned the maths behind it.

The parts of the brain that make this possible can be neatly divided into three. One part deals with quantities smaller than four. It means we can immediately see the difference between one apple and two. Another part is used for larger quantities, and the third recognises geometric shapes. That is how people who have never seen a map before can use one to work out the way from A to B.

We can all deal easily with small numbers, even as babies. We are born with the ability to distinguish between one and two. Not, of course, between the numbers, but between one thing and two things. Babies are surprised, for example, if they spend some time looking at a piece of paper with a single dot on it and suddenly see one with two dots. Their surprise shows that they understand that they are seeing something else. Scientists can establish that by measuring how long a baby looks at the piece of paper. The baby will soon get bored looking at the same image, but will look at a different one a little longer.

This allows researchers to study in detail what babies expect from the world around them. And it leads to surprising discoveries. Babies seem, for example, to be able to add and subtract. If you show a baby two dolls and then take one away, the baby will expect there to be only one left. If you then start with two dolls, take one away and there are still two left, the baby is very surprised. Before babies have learned anything about numbers, they apparently understand that $2 - 1 = 1$ and not 2!

That is, of course, not strictly true. We now know that what surprises the babies is that a doll has suddenly appeared that they knew nothing about. The same happens if they see $1 + 1 = 1$. They are then surprised because a doll has disappeared without them being aware of it. This is because part of our brain is specialised in keeping track of the things around us: what colour they are, how big, and so on. We record that kind of information when we focus on something. Babies do that too, so they notice it immediately if something suddenly disappears or appears in a place where they are certain there wasn't anything before.

Our brain can only keep track of a small number of things in such detail. For babies, the maximum is three: any more than that and they get confused. There is an experiment in which babies have to choose between two things. To their left is a box with one biscuit in. They saw the biscuit being put in the box, so they know it's in there. To their right is a box with four biscuits in, which the babies also saw being put in. So which box will they choose? Which one will they crawl to?

Strangely enough, they don't always choose the box on the right. You would think that babies that can tell the difference between one biscuit and three would also be able to distinguish between one and four. As the difference is larger, the task should get simpler. But no: if there are four biscuits in the right-hand box, they have no idea which box contains the most biscuits and they choose entirely at random. The useful part of the brain that can distinguish small amounts goes into overload and gives up. For the first twenty-two months of their lives, children can't tell the difference between one and four.

That breakthrough around twenty-two months does not happen because the brain can suddenly do four things at the same time. Adults may be able to do that, but even they will find it a challenge to track four objects simultaneously. We still don't know exactly what happens, but it has something to do with language. Children learn to see the difference between one and four more quickly if they speak a language that distinguishes between singular and plural. Children in Japan, for example, take longer to learn the difference because Japanese doesn't make that distinction. They catch up again later, though, as children who speak languages like Dutch or German take longer to learn numbers bigger than 10. In Germanic languages, 24 is spoken as 'four-and-twenty', while in Japanese – as in English – it is 'twenty and four', making it easier for kids to understand how numbers progress as they get bigger. In French it is even more difficult: the word for ninety is 'four-times-twenty-plus-ten'.

Language is therefore important in learning numbers, but in the end the most important thing is the ability to see the difference between one thing and more than one thing. That is probably the basis on which children learn what the word 'one' means. Before they can do that, children don't know how numbers work. They can reel them off – 1, 2, 3 etc. – but if you ask them to give you one toy they will give you a random number of toys, no matter how often you count off the numbers with them.

This is how we build on the things we are born knowing. By learning what 'one' means, we can also learn that 'two' means 'one and another one'. That is ultimately possible only because of the parts of our brain that work with small quantities – incredibly convenient, especially when learning exact numbers. Among the cultures we looked at earlier in this chapter, the part of the brain that deals with larger quantities is more important.

I don't know exactly! The brain and large numbers.

As soon as we have to deal with more than three things, a different part of the brain kicks in, one that also works from birth. Babies can immediately tell the difference between four and eight dots. But – and that's the difference between this part of the brain and the part that deals with smaller numbers – we can't do it for all larger numbers. Babies can't, for example, see the difference between four and six dots. They do know, however, that sixteen dots are more than eight. That's because, once there are more than three or four dots, we can't tell exactly how many there are. We can distinguish differences between some things and not others. Newborn babies can see if there are at least twice as many dots on one piece of paper as on another. So they can't tell that six is more than four, but do know that eight is more than four. It depends on the relative difference between the two numbers. Think about it: it's much more difficult to see the difference between 100 and 105 at a glance than between five and ten.

We can see more and more differences as we get older. By the time they are a few months old, babies can distinguish between four and six dots, a difference that is only one and a half times as big. Generally speaking, adults can even see that thirteen dots are more than twelve. They're not always right, but generally they can tell which is the larger number. On the other hand, it is almost impossible to tell the difference between twenty and twenty-one without counting.

In the end, then, counting works better. Unlike what we can see without counting, numbers are exact. That's why the Loboda don't know precisely how many yams they have given someone. They can see roughly how many there are and can tell immediately if someone gives them a lot fewer (or more) back. But no one will notice one more or less.

We thus deal with larger numbers differently than with small ones, even though babies sometimes seem to be able to cope with large quantities before they know anything about numbers. An experiment with dolls to test their reactions to 2 − 1 = 2 can also be used with 5 + 5 = 5. They know that is wrong and will react with surprise, while they find 5 + 5 = 10 completely normal. Does that mean we've already taught them to count with larger numbers?

Again, that was the conclusion of the researchers who conducted the experiment in 2004; but, here too, we now know better. While babies respond with surprise to 5 + 5 = 5, they find 5 + 5 = 9 just as uninteresting as 5 + 5 = 10, simply because they can't tell the difference between 9 and 10. They are therefore surprised when they only see five dolls because they expected more. But they didn't expect exactly ten, which they would have done if they had worked it out mathematically. What they expected was something much less accurate: more than five but not a lot more. Addition and subtraction are unfortunately something we all have to learn.

How does all this work? What does the brain do to allow us to see such differences? The jury is still out on these questions. Before I tell you what I think about it, I want to explain about the part of our brain that deals with length, time and suchlike.

Length is also something we can't see exactly just by looking. Of course, we can see immediately if one thing is roughly twice as long as another. I can see the difference between the long and short sides of a rectangular table, but not in exact centimetres. I can make a wild guess, but the chances are that I'll be wide of the mark. The same applies to time: I have a reasonable idea how long something takes. I know the difference between ten seconds and five minutes, and between one and two hours. But I won't notice the difference between an hour and an hour and one minute.

Does that sound familiar? It probably does, because the

way we deal with length and time is very similar to how we work with quantities. Babies can also see the difference between lengths from birth. They can even tell that the time between two sounds is longer or shorter, as long as the times are sufficiently different. We see more and more of these differences as we get older and we become better at recognising lengths and periods of time. But without measuring, we don't know exactly how long they are. With every bridge the Kewabi build, it is trial or error whether the tree trunks are the right length. And the greater the distance to be bridged, the greater the chance that they will get it wrong, simply because they don't know how far it is until they try to lay the trunk across the river.

We all possess the powers of estimation of these tribes that don't measure or count. We are born with it and it gradually gets better. We also share them with other species: primates, rats and goldfish, to name a few, can all see the difference between quantities and lengths. Almost every species has a part of the brain that deals with these things. How is that possible? What enables us to work with quantities without any knowledge of maths? How can other animals do that, too?

My answer is that because we can work with lengths and times, our brains use that information to tell us something about quantities. The fact that we can easily understand lengths and other visual phenomena enables our brains to use that ability to learn about more abstract things. On the basis of what we see in terms of lengths, surface area, etc., our brains start abstracting and come up with quantities.

Why do I think that? Partly because we can play all kinds of tricks on our brains. We can see that the part of the brain used for lengths and so on is the basis for its grasp of quantities, since misleading information about lengths will also influence our estimations of number. And that is exactly what we should expect when our brains learn about quantities by abstracting from length, density, timing, etc.

The most convincing example of this is the illustration below. Take a quick look at it and try to tell, without counting or thinking about it too long, which of the circles contains the most dots. There is a good chance that, like me, you will choose the circle on the right. That one seems more crowded so will probably contain more dots. But that impression is misleading: count them and you will see that each circle contains the same number of dots.

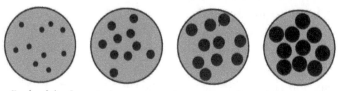

Each of the four circles contains the same number of dots, but the bigger they are, the more there seem to be

These are the kinds of mistakes our brains make and, because they clearly show where things can go wrong, they tell us something about how the brain works. There are other ways in which it doesn't work optimally. If we have to decide whether one number is larger or smaller than another, for example, it helps if the number is on the 'correct' side. Our brains expect smaller numbers to be on the left and larger ones on the right, and we are more likely to give the right answer if the numbers are presented in that way. If we are asked whether 9 is bigger than 5, we are likely to answer more quickly if the 9 is to the right of the 5. The difference in thinking time when the 9 is on the left or on the right is not long enough to notice yourself, but a timer will pick it up. If, of course, the 9 is the smaller number – is 9 bigger than 15? – it is easier to answer if it is on the left.

This doesn't apply to everyone: for people who speak Hebrew, for instance, the opposite is true. As Hebrew is read from right to left, it's easier to guess the smallest number if it's on the right. It can be even more confusing for people who speak two languages fluently. For someone who speaks both Hebrew (right to left) and Russian (left to right), it depends on which language they last read. If it was Hebrew, it is easier if the larger numbers are on the left; if it was Russian, the brain prefers to see the bigger numbers on the right.

In other words, our brains link numbers to what we see. The location of a number determines how the brain deals with it, and not only if it is an exact number like 9 or 15, but also if it is a bunch of dots. What's more, we are not the only animals that do this. Chicks also prefer to see larger numbers, in the form of dots, on the right. Reason enough to suggest that, for chicks too, their ability to estimate numbers comes from their grasp of lengths.

Recognising geometric shapes – even chicks can do it

We therefore know quite a lot about how people can engage in all kinds of activities that involve numbers, like trading and building bridges or seagoing canoes, without using maths. But another branch of mathematics is also of great importance to human society: geometry. We need to have some understanding of geometric shapes to build houses: how the surface area will be affected if we make the house longer, or what happens if we change the radius of a circle. Fortunately, we also have inborn geometric skills that enable us to do all that.

There is even a specific part of our brains that deals with shape, especially to make sure that we can find our way. Again, not only people have that ability: other animals, including chicks, can also recognise simple shapes and use them, for example, to find hidden food. This is of course not the only way animals find their way around: migrating birds navigate by the Sun and the stars, and insects use scent spoors to get back to their nests. Awareness of shape is by no means always necessary, but it can be useful – for example, if a nest is in the middle of a circle or in the corner of a rectangle.

Researchers can easily replicate such situations to test how well animals and children can recognise geometric shapes. They hide a small amount of food in a space with a certain geometric shape and watch where the test subject searches for it. The illustration on page 51 shows an experiment in a rectangular space. A chick starts in the middle of the rectangle and has to find a treat in one of the corners, which the researcher buried there while the chick was looking at it. To make it more difficult to locate the treat, the chick is spun round very fast before it starts to search. The chick seems to know that the treat was in one of the corners where, seen from the middle, the long side was to the left of the corner. After being spun round, it only looks in two places: the lower left and the upper right. One of these is the correct location of the treat and the other is its mirror image. This is a perfect result, because it is not possible to choose between the two. Both corners have the longer side of the rectangle on the left and are otherwise identical for the disoriented chick. It does show, however, that the chick recognised and retained the information about the longer side being on the left.

Sometimes the chick is unable to recognise the rectangle.

In the two lower pictures, it isn't aware that it is standing in a rectangle and looks for the treat in all four corners. Although chicks are quite good at recognising figures, they are sometimes not clear enough.

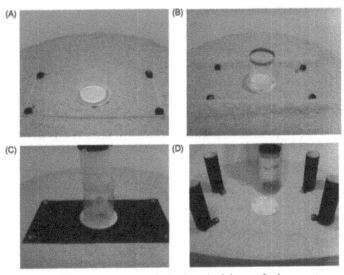

Four different rectangles in which a chick has to find a treat*

Other animals can also recognise geometric shapes, with roughly the same limitations. Rats, pigeons, fish and rhesus monkeys have all shown that they can distinguish different shapes. And, with our own species, small children can spontaneously deal with shapes. A young child looking for a sweet in a rectangular room will search in only two places: the right place and its mirror image. They will continue to do so even when given extra information. It makes

* Elizabeth S. Spelke, *Space, Time and Number in the Brain: Searching for the Foundations of Mathematical Thought*, chapter 18, Natural Number and Natural Geometry, figure 18.5 , reprinted with permission from Elsevier

no difference, for example, if the wall where the sweet is hidden is a different colour. They only learn to use such information at a later age.

The question now is: do these experiments actually prove that children and animals can recognise shapes? Do they know what a rectangle is? Or do they only know that something is in the corner with the long wall to the left? Further research has shown that they really do take account of shapes and not just corners and lengths.

When I am in my office, for example, my brain actively registers an image of the room. I remember not only that my desk is in a corner with a long wall to the left, but also that behind me there is another desk on the left and a door to the right, and so on. Of course, I am very familiar with this room; after all, I work in it almost every day. But whenever we go into a new room, too, our brains make a mental picture of the whole room. Even with a blindfold, you can describe the general layout and specify more noticeable objects.

What you don't remember is exactly where you are in the room. Imagine you are blindfolded and someone spins you round quickly so you don't know which way you are looking. Of course, you can no longer point out where the various objects in the room are. Even if a light is on and the blindfold is thin enough to let some light through, you won't be able to say where things are. But you will still be able to describe the layout of the room. You have the mental picture of the space you are in, but it is completely impossible for your brain to tell you where you are.

To make that mental picture, you have to have some ability to recognise shapes: the number of corners, where the walls are in relation to each other, etc. Although this example refers to adults, there is plenty of evidence that the same applies to children and animals. The brain even proves to have separate neurons for squares, circles, and so on.

It is thanks to those neurons that people in cultures that live without maths can deal with geometric shapes. Like the Pirahã, the Mundurukú – who live in various parts of the Amazon rainforest – don't use maths and rely entirely on their inherent skills. The Mundurukú have taken part in experiments with shapes. A member of the tribe was given a piece of paper showing six geometric figures, one of which was different from the rest: for example, five straight lines and one crooked one. The question was whether people without mathematical training can recognise the difference, in the same way that we looked earlier at whether children can tell the difference between one biscuit and four. Sometimes the test subjects easily spotted the difference – for example, between a straight and a crooked line. At other times, such as between a dot in the middle or elsewhere on a line, they found it more difficult.

Even though they don't always get it right, these people don't need extra lessons to understand shapes and distances. They understand them well enough to read a map without instructions, at least a simple one showing only a small area. In one experiment, a Mundurukú woman was asked to walk to a cylinder in a field after studying a simple map. The map is shown in the illustration overleaf. It is a rectangle containing three geometric figures, one of which is a different colour to the other two. After looking at the map, the woman walked to the coloured cylinder in the field. This showed that she understood that the map represented the field. She also completed the test successfully without the help of different colours.

This ability to read maps has its limitations: some shapes are more difficult to recognise than others, and finding the right place is more tricky if the shape isn't a different colour. What's more, the road maps we use are slightly more abstract and look less like the area they represent than the maps used in these experiments. But – and this is what concerns us here – the experiments do show that we are able to understand geometric shapes and figures without mathematical training.

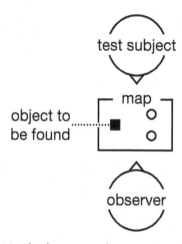

Mundurukú map-reading experiment

Does maths add anything extra?

All kinds of cultures are able to live perfectly well without figures and geometry. That's because, from birth, we know how to deal with quantities, distances and shapes. Our brains are designed so that we don't need maths to know roughly how many yams there are in a basket, how far it is across a river or how many trees we need to build a house.

Yet we mustn't confuse these inherent skills with mathematical skills. Maths is something we have to learn. Babies don't know what numbers or geometry are. They recognise shapes but they don't think about them or analyse them. Thinking about and analysing shapes is what mathematicians do. Simply being able to recognise a shape isn't the same as doing maths.

So why bother with maths at all? It is becoming increasingly clear that we don't need it to survive. We can even lead very happy lives without ever learning anything about it. And yet people around the world, from Mesopotamia to Egypt and from Greece to China, have found it necessary to delve into the world of mathematics. It added something extra that was very important, something they couldn't do without. In the next chapter we will look more closely at exactly what that was.

CHAPTER 4

Maths long, long ago

An overseer near Umma, now a ruined city in the south-east of present-day Iraq, draws up his annual report. It is 2034 BC and King Shu-Sin rules over the whole region. The overseer has a serious problem. Every year the state specifies how many days his gang has to work, and every year he comes up short. Over the years he has built up a deficit of 6,760 working days, and this year, partly because of a miscalculation on his part, that has risen further to 7,421.

In those times, working days were seen as goods that belonged to the state. As far as King Shu-Sin is concerned, it is the overseer's fault that not enough grain, etc. has been produced. When the overseer dies, his house, his possessions and his family will be sold to pay off his debt to the state.

It was a tough life, both for overseers and for the workers in their gangs. Women only had a day off once every six days and fit men one day in ten. No one retired; old people carried on working, probably until they dropped dead. How did King Shu-Sin keep this system going? By bookkeeping. The overseer's annual report was a proper report, just like those that companies produce today. Shu-Sin supervised his state through a system of double-entry bookkeeping with credit and debit balances, verified by receipts, vouchers and promissory notes. Shu-Sin's bookkeeping was so comprehensive that, after his rule, nothing like it reappeared until 3,500 years later, around 1500 AD in Europe. And it was even longer before a state decided to set up a similar centrally planned economy.

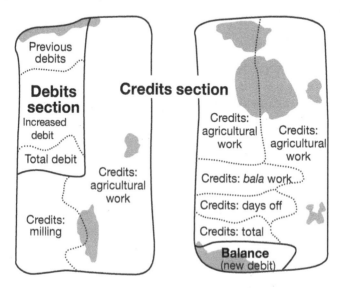

Bookkeeping in 2000 BC

The whole system was horrendous. The number of required working days was so absurdly high that virtually all overseers were in debt. The only advantage of the system was that it left an enormous amount of clay tablets behind for us to find. All those receipts, invoices and annual reports have been preserved, which is why we know so much about that overseer in Umma. His annual report for the year 2034 BC has been unearthed and, except for a few grey spots here and there, is perfectly readable.

The report also shows us what numbers are good for: bookkeeping. It is much easier to plan and keep track of working days if you can work with exact quantities. Mathematics makes it easier to manage big groups of people, and that's why maths didn't develop until people started living together in cities in large numbers.

Pushing the envelope

Hunter-gatherers lived in Mesopotamia, roughly present-day Iraq, long before King Shu-Sin. They built the first settlements in the area as early as 8000 BC and started to grow grain, vegetables and fruit. It was a great success: two large rivers and a clever irrigation system enabled them to feed larger and larger groups of people. Cities emerged and had increasing contact with each other, with merchants travelling back and forth in the hope of earning a lot of money. It became increasingly important to have some form of central authority. People who lived in tribes found it easy to maintain law and order because everyone knew each other, but that was no longer possible once cities had grown too large.

Governments evolved in the form of city states, and started to impose taxes. That didn't go so well at first, because there were as yet no numbers to work with. It was like the Loboda and their gifts. The state didn't always take the same amount; it just guessed what it needed. So there was no way of knowing how much you would have left after paying your taxes or of checking whether you paid the same tax rate every year. What made it even more complicated was that there were hardly any words to tell people how much they had to pay. A simple description like 'a basket' already implies a number: one. Talking about quantities was difficult when there were as yet no numbers. But the city states came up with a solution.

It all started with food stores. In the cities of Susa and Uruk, both in Mesopotamia, the stores became bigger and bigger as the cities grew in size. To keep track of how much food was in them, merchants started to use small clay tokens, all the same size, with markings on them. Each token represented a quantity of food: a basket of grain, for example, or a sheep. That meant that you no longer needed to see all the separate baskets or sheep in a store; you could tell from the number of tokens.

The tokens were used for more and more things. When the taxmen in Susa had to tell the people of the city how many baskets of grain were needed, they had a problem because there were as yet no words for numbers. So they used sealed clay envelopes containing tokens. Every token represented a basket and everything went fine, without the need to count. People in Susa were already using tokens around 4000 BC to manage payments to the temple and impose taxes, though they didn't of course work out afterwards how much they had collected, as they had no numbers with which to do that.

In Uruk, the authorities went a step further. As in Susa, they started using tokens in clay envelopes to let people know how many goods had been sent or how many they had to send back. The envelopes were also a good way of ensuring that nothing was stolen en route. But they found a clay envelope filled with pieces of clay a little cumbersome. We don't know exactly when and how, but at some point someone came up with the idea of making drawings of the tokens on the outside of the envelope. It wasn't easy to rub out markings on a clay tablet, so a drawing on the outside of the envelope was just as safe as loose pieces of clay inside. The markings gradually evolved into numbers. People forgot what they stood for and started seeing them more as symbols for a basket of grain, a sheep, etc. These were the first written words, much earlier than other words: full sentences didn't appear on clay tablets for another 700 years.

That's how the first numbers evolved in Mesopotamia. Tokens were portrayed on the outside of clay envelopes, which were then replaced by flat clay tablets. The symbols were gradually used more generally and, because drawing the same symbol over and over again is a lot of work, new ones were devised for repetitions. And that led to numbers: if you use the same symbol to count ships and grain, you are using a number. And all because cities like Susa and Uruk became so big that they needed a convenient way to collect taxes.

The first numbers looked like cones and circles. That's because the pens they used to write on clay tablets had two sides: the back was rounded and the front more pointed. Pressing the back into the clay produced a circle, while the front left a cone. These number symbols started on the right, with a small figure like a rounded cone on its side representing 1. The small cones were repeated until there were nine of them lined up next to each other. The 10 was a different symbol, a small circle.

The first numbers in Mesopotamia

The Mesopotamians didn't continue counting in series of tens like we do. The small circle was replaced after six times, i.e. after 59, with a larger rounded cone representing 60. And so it went on, in alternating series of ten and six symbols, until they reached 36,000. Anything bigger than that was difficult, but that didn't matter: who had 36,000 baskets of grain in a store in those days?

The sexagesimal (60-based) number system in cuneiform script

Later, when the Mesopotamians had developed a more complex script, they also wrote numbers differently so that they could record bigger quantities. These numbers are shown in the illustration above. That script is today called cuneiform, which means 'wedge-shaped', after the characteristic form of the symbols. They could even write fractions using this method – all to keep the economy going.

City states like Uruk and Susa used numbers not only to collect taxes but also to manage the supply of food. They kept a record of how much grain and other food was in the stores and how much was still in the fields, and whether it would be enough to feed the whole population. They estimated how much they needed to produce bread for everybody and, if there was not enough, they planted more. This, too, needed to be planned: too much food was almost as bad as not enough, since it would be ruined if it lay in the stores for too long.

The bookkeepers were the best scribes in Mesopotamia

and, together with the temple priests, were responsible for all that planning. Scribes learned not only to write but also to count and to measure: as well as keeping the books, they could measure out the area of a piece of land. They also drew up contracts for merchants, and some worked out how many workers would be needed for a building project. Maths was used to plan more and more activities and buildings were designed using geometric figures. Scribes became architects, and ended up as overseers for King Shu-Sin.

Schoolwork in Mesopotamia

Of course, the scribes needed to be trained for all these different tasks and, thanks to the excavation of a school from 1740 BC, we have a reasonably good idea of how that happened and what was considered important. It wasn't just about performing basic calculations in your head or dividing land into plots. The lessons also devoted a lot of attention to applying maths to everyday problems. That's what scribes were educated to do and, as a satirical text found in the school testifies, anyone who had no idea of how to apply maths practically was a target for ridicule.

The text relates how a young scribe is talking to an older, more experienced colleague. The old man complains that the standard of teaching has deteriorated badly. Today's youngsters can't do anything any more, not even divide up a piece of land between two people. The young scribe disagrees and insists that of course he can divide a piece of land into two. He tells the old man to take to him to a random spot and he will show him how well he can do it. Chuckling, the old man tries to explain that he doesn't mean using ropes to divide up the land. He means calculating it – that's what you need for a contract and the foolish young man clearly can't do that.

Mathematics was intended to be of practical use but, for a long time, the link between the sums and the reality was not taught explicitly. A large part of the teaching at the school – in Nippur, another city in central Mesopotamia – entailed the pupils repeating rows and rows of sums over and over again. If you copy what the teacher does often enough, you will eventually learn how to do it yourself. That applied as much to maths as to any other subject.

Pupils in Nippur first learned to read and write, of course, especially by writing out lists of words repeatedly until they knew them by heart. After learning the words for places, sorts of meat, weights, lengths, and so on, they started doing maths. That also meant memorising multiplication tables and other lists of facts about arithmetic and geometry. As the cherry on the cake, they also learned a few standard contracts off by heart by – you guessed it – writing them out many, many times. But not everything was a matter of repetition. Sometimes pupils were given mathematical word problems to solve based on practical situations.

A wall. The width is [2 cubits], the length is 2 ½ nindan, the height is 1 ½ nindan. [How many bricks are in the wall?]
A wall. The length is 2 ½ nindan, the height is 1 ½ nindan, the bricks are 45 sar_b. How thick is the wall?
The surface area of a house is 5 sar_a. How many bricks do I need to build the walls to a height of 2 ½ nindan?

It's logical to want to know things like this, but the pupils were also given a lot of nonsensical word problems, such as:

A wall. The height is 11 nindan and the bricks are 45 sar$_b$. The wall is 2.20 (our 140: 2 × 60 + 20) nindan longer than it is wide. What is the length and width of the wall?

[A wall] of baked [bricks]. The height of the wall is 1 nindan, the bricks are 9 sar$_b$. The sum of the length and thickness of [the wall] is 2.10 (our 130). What is the length and thickness of the wall?

[A wall of baked bricks]. I lay out 9 sar$_b$ of [baked] bricks. The wall is 1.50 (110) [longer than it is thick]. The height is 1 nindan. [How] long and thick is the wall?

Think about the first and third problems: you have a wall and you know exactly how much longer than wide or thick it is. But how do you know that? Not by measuring it, clearly, otherwise you would already know the answer. But how else would you have that information? It all sounds a little far-fetched. The second problem is even more bizarre: how can you know the sum of the length and the thickness of the wall without knowing the length and thickness? Yet another problem you'll never have to solve in practice.

The idea behind such nonsensical mathematical word problems was, however, not to show how well maths could be used in everyday situations. They were probably intended to test how good the trainee scribes were at maths. The only disadvantage was that this more complex mathematics is no longer as useful. By improving its pupils' mathematical techniques, the school moved further away from the practical reasons for using maths in the first place. Such calculations have nothing to do with organising a city state.

And yet it's not so crazy to learn about this kind of maths. After all, you never know when it might produce something useful. Look at the illustration opposite, which shows a stick leaning against a wall. Assume the stick is five metres long (c) and touches the wall at a height of four metres (b); how far from the wall does it touch the ground (a)?

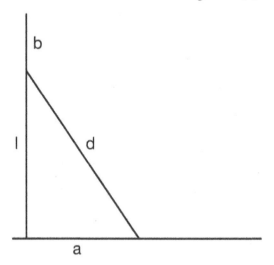

Pythagoras' theorem, but back in Mesopotamia

If you still remember anything about triangles, you already know the answer. The stick and the wall form a right-angled triangle, which means the lengths of the sides can be calculated using Pythagoras' theorem: $a^2 + b^2 = c^2$. Since we know that b is four metres and c is five metres, we can work out distance a (from the base of the wall to the stick): $a^2 + 4^2 = 5^2$ -> $a^2 = 5^2$ (i.e. 25) – 4^2 (i.e. 16) -> $a^2 = 9$ -> $a = 3$.

What is amazing is that the Mesopotamians knew that too, 1,500 years before Pythagoras was born. You probably wouldn't use his theorem for a stick against a wall – it would be easier just to measure the three lengths. But it is a very useful way of making sure you have a right angle. If the three sides of the triangle match up to $a^2 + b^2 = c^2$, you have a right angle.

Mathematics was very advanced in Mesopotamia. They could already solve a whole range of difficult problems around 1800 BC, much earlier than the Greeks. They could, for example, solve the formula $x^2 + 4x = {}^{41}/_{60} + {}^{40}/_{3600}$ ($x = \frac{1}{6}$ being one solution). At least, as long as you didn't live in the time of King Shu-Sin, who was concerned that maths encouraged people to think independently. He banned complex maths from the curriculum during his reign, which left plenty of time to indoctrinate the children to be loyal subjects.

To return to the question I asked in the previous chapter: why did people start using mathematics in the first place? In Mesopotamia, it was needed to organise city states. Maths made it easier to collect taxes, plan food supplies and build houses. With so many people, it had become very difficult to do these things without it. But not all maths was useful. Solving mathematical word problems with no practical value was a status symbol, to show others how clever you were. Even King Shu-Sin did them: his subjects weren't allowed to know anything, but he of course knew everything.

Bread, beer and numbers in Egypt

Two men in ancient Egypt are thinking about what profession they should choose. One says he'd like to be a farmer, but the other one says: 'No! A scribe, that's a good way to make a living. A farmer has to work hard all day, ploughing the land, getting in the harvest, keeping the irrigation system going, etc. All a scribe has to do is sit somewhere warm and write things down.' 'OK,' the first man says. 'A farmer is a bad idea. How about a builder?'

You can guess the rest. In this satirical text, all kinds of professions are compared to each other and every time it's the scribe who comes out doing the least work. The moral of the story is clear: 'The scribe calculates the taxes that everyone has to pay. You won't forget him.' Today, this difference in work levels is less marked, but our tax services still use a lot of maths.

Ancient Egypt was very similar to Mesopotamia. There, too, mathematicians – the scribes – had an important role to play in collecting taxes. But with one big difference: we know a lot less about ancient Egypt. Everyone in Mesopotamia wrote on clay tablets, which we have found practically undamaged. But in Egypt they wrote on papyrus, which perishes more quickly. What's more, the places the Egyptians lived in are still cities, like Cairo and Alexandria, which makes it harder to dig up the historical remains that lie beneath them. That's why there are only six texts on maths from ancient Egypt, and they all date from the Middle Kingdom (2055–1650 BC). We know much less about the Old Kingdom (2686–2160 BC), when the Great Pyramid of Giza was built, and the New Kingdom (1550–1069 BC).

But didn't the Egyptians write hieroglyphs on stone, which tends to be well preserved? That's true, but they only tell stories about kings and gods. They used an entirely different kind of writing, known as hieratic script, to keep all their administrative records, and that script did contain numbers.

Those numbers first appear in written documents around 3200 BC, about the time the first clay tablets were being used in Mesopotamia. In Egypt also the first documents were administrative: lists of people, places and goods, with quantities. Some even recorded the water level in the River Nile, probably to help in calculating taxes. So there, too, numbers were first used to collect taxes and take stock of how much food was available twice a year.

1	2	3	4	5	6
7	8	9	10	20	30
40	50	60	70	80	90
100	200	300	400	500	600
700	800	900	1000	2000	3000
4000	5000	6000	7000	8000	9000

Numbers in hieratic script in ancient Egypt

As the illustration above shows, the Egyptian system of numbers looked a little like ours, with a new symbol after 9 and another after 99, etc. The only difference is that they didn't have a symbol for zero (0): that didn't come until much later, in India.

The Egyptians also had symbols for fractions, which were indicated by a dot above the number. A number 2 with a dot above it is ½. Today, to make it easier to read, we use a short line above the number, for example $\bar{2}$.

For the Egyptians, fractions were the opposite of a whole number (½ is the opposite of 2). But a fraction like ⁵⁄₇ is not the opposite of 7 or another whole number. Yet such fractions did come up, if only in the administrative records.

They needed a clever solution so they could work with these more complicated fractions, and so they devised the idea of writing them down as the sum of fractions that could all be divided by 1: ¾, for example, could be written as ½ + ¼, or $\overline{2}\overline{4}$. It was even possible with ⁵⁄₇: ½ + ⅐ + ¹⁄₁₄, or $\overline{2}\,\overline{7}\,\overline{14}$. Try it for yourself for a different fraction, just to see how difficult it is. That's why the Egyptians learned the most important fractions off by heart.

They used these fractions a lot in their records, especially for taking stock of bread and beer, the core products of their economy. There was as yet no real money. Egypt didn't use coins until 390 BC, when it started recruiting Greeks for its army: the Greek soldiers flatly refused to be paid in bread and beer, demanding their wages in the form of silver Greek coins. So the Egyptians started using the coins and discovered that money was actually quite useful.

Before the Greeks arrived, the Egyptians had kept their economy afloat for thousands of years without using money. They built the pyramids without it, though the stories of large gangs of slaves are a myth: the people who worked on the pyramids were regular paid labourers. What is surprising is that they were paid only in bread and beer, sometimes in quantities expressed in fractions. Whole payrolls have been found, showing that even priests were paid in this way, getting for example $\overline{2}\,\overline{3}\,\overline{10}$ (2 ²³⁄₃₀; the one exception to the 1/... rule is 3, which could be written as ⅔ with a double line) vats a day. They weren't intended to drink it all themselves, but bartered what was left for other goods.

That's how it went with everything. If you needed a bed, you exchanged other goods to pay for the bed you liked. You even bought a house by exchanging goods with the previous owner. But barter in Egypt wasn't as hit and miss as it was with the Pirahã, because the Egyptians could count.

Prices for goods were stable and, for bigger purchases like a house or an ox, they would go to a scribe. He would draw up a contract describing the exchange, so that none of the parties could complain afterwards. And that's why scribes had a lot to do with bread and beer: payslips and contracts were full of them.

The army needed food too, so there was a scribe to ensure there were enough supplies. One of the texts from the Middle Kingdom is about a scribe who pokes fun at a colleague by describing situations in which he would make mistakes. The colleague estimates, for example, how much food will be needed to feed an army unit of 5,000 men on a long campaign. He concludes that they will have to take with them 300 loaves of bread and 1,800 goats.

On the first day of the campaign the army arrives at the camp, where the scribe has all the provisions waiting for the men. Proudly, he shows them everything he has bought for them and they immediately start eating, wanting to store up a lot of energy for the long day's march ahead. After an hour, they've eaten all the food. So they go to the scribe, complaining loudly: 'The food's all gone! How's that possible, you idiot?!' The scribe has no answer for them and has to offer his resignation.

Scribes were like managers. Because they had the unique skill of being able to count, they were responsible for wages, taxes and rations. They also calculated the size of areas of land before and after the Nile flooded, so that farmers could get compensation for land they had lost. They even calculated how long it took for someone to make a pair of shoes, so that it could be coordinated as closely as possible with the supply of leather.

More impressive than all these practical applications of mathematics is that the Egyptians used maths to build the pyramids. When you build a pyramid, you have to know what angle you need in order to end up with a sharp point at the top. As you have to start at the bottom, you can't estimate that. But you can calculate it – and that's what the Egyptians did.

You need certain details to do that: how long and wide the pyramid is to be, and how high it will be when it's finished. If all four sides aren't built at the same angle you won't end up with a sharp point. That makes the angle very important, since it dictates the height and appearance of the pyramid. The Egyptians didn't, however, work with angles in the same way that we do. They didn't talk about degrees; they had a different method.

You can measure an angle just as easily by looking at how much the side of the pyramid shifts as it goes up. This is known as rise (vertical shift) and run (horizontal shift). The illustration below shows how that works. If the angle is 90 degrees, the sides of the pyramid will go straight up and the run will be zero. The smaller the angle, the bigger the run will be. At an angle of 45 degrees, the rise and run will be the same. In other words, the sides of the pyramid will move as far vertically as horizontally.

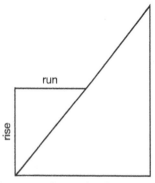

Measuring the angle of a pyramid

All in all, the Egyptians used maths for a lot of things. Although we don't know so much about them because of the short shelf life of their writing materials, enough papyrus has survived for us to see that they actually used the same mathematical methods as the Mesopotamians. Mathematicians were important and mainly worked as record-keepers. Egypt had a clever tax system that took account of the flooding of the Nile, and there were standard contracts for bulk buying and selling of goods. Later, around 300 BC, the Egyptians adopted somewhat more complex mathematics from Mesopotamia, but what they did until then they had worked out for themselves. They are another example of a people that 'simply' used complex mathematics.

The ever-theoretical Greeks

No one in antiquity knew as much about maths as the Greeks. That's why there are so many renowned Greek mathematicians, the most famous of all being Pythagoras, Euclid and Archimedes. Surprisingly enough, we don't know that much about how the Greeks used mathematics. Stories and texts from ancient Greece have survived, but they are all theoretical. Euclid, for example, is well known for his book on theoretical geometry, which contains all kinds of definitions and proofs, like 'a line is length without width'. It is abstract, along the lines of Plato, but doesn't tell us how maths was used in practice. The same applies to all the other theoretical works: they offer little insight into how the Greeks practised maths, why they started doing it and what reasons they had for writing down all those abstract theories.

That is not to say that the Greeks didn't apply their theories. An impressive example is the Tunnel of Eupalinos on the island of Samos. For 1,200 years the tunnel – more than a kilometre long and less than two metres wide – carried five litres of fresh water a second to the capital of the island. The Greeks not only managed to dig this tunnel in 550 BC but, even more remarkably, they started at both ends at the same time. In some way or other they successfully joined up the two halves of the tunnel in the middle. A couple of metres either way and the two digging crews would have passed each other underground.

We don't know exactly how they did that, since the Greeks apparently didn't think it was worth explaining – unlike the Romans, who found mathematical applications like this very interesting. The Greeks probably used straight lines and right-angled triangles to make repeated measurements and adjust the direction of the digging. In the middle, when the two halves were very close together, they could hear each other's hammers and followed the sound to break through the final metres. Using a combination of continual measurements and ingenuity, they succeeded in digging a tunnel a kilometre long. And they made such a good job of it that we can still visit it. It could probably even be used as an aqueduct again if necessary.

It isn't easy to find out how the Greeks applied their theoretical knowledge in practice. Despite their advanced theoretical achievements, a lot of it was guesswork. Pythagoras may not have devised the theorem named after him – the Mesopotamians knew about it much earlier – but he was the first to prove it in the same way that modern mathematicians provide evidence for their theories. Neat, logical mathematical reasoning showing that what they are saying is completely watertight – that's what the Greeks are known for. Euclid had his book of proofs and Pythagoras had his theorem. Archimedes also proved a number of theorems, but was known for many other accomplishments of his own. Good reason for him to stand head and shoulders above the rest.

Archimedes was the physicist who allegedly discovered his famed principle of fluid dynamics while taking a bath. According to legend, he was so excited that he ran and told the king without getting dressed first. Archimedes was also apparently such a good designer of war machines that, for many years, the Romans dared not attack his home city of Syracuse. The threat of Archimedes was enough to deter them. When the city was finally captured, Roman soldiers were sent to his house to take him captive. Archimedes was busy contemplating a mathematical problem and said to them, 'Don't disturb my circles!' The soldiers killed him, much to the displeasure of their superiors. Whether all this is true, we will never know. There are many more strange stories about Greek mathematicians. Pythagoras is alleged to have thrown a pupil overboard, causing him to drown, to keep secret his proof that you can't write every number as a fraction.

All the stories aside, we know that Archimedes was a brilliant mathematician and was especially good at thinking about volumes and surfaces. His tomb bore images of a sphere and a cylinder in commemoration of this most famous mathematical discovery: he was the first to show how the volumes of a sphere, a cylinder and a cone are related. For the Greeks, working out the volumes of geometric objects was a tough problem, as they had no formulas to go by. One particularly difficult puzzle was finding a square with the same surface area as a circle. This is still reflected in the expression 'squaring the circle', for doing something that is impossible.

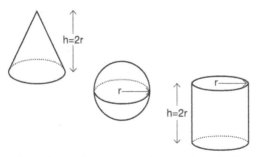

The volume of a cone, a sphere and a cylinder

Archimedes showed how much bigger a cylinder is than a sphere or a cone with the same radius. In the three figures in the illustration, the radius is shown with the line r. The height (h) of the cone and the cylinder is twice that of the radius (2r), and is therefore the same as the diameter of the sphere. If you cut away the correct pieces of the cylinder you will end up with the cone and, in a similar way, the sphere also fits neatly into the cylinder. Reason enough to assume – correctly – that the volumes are related: the volume of the sphere is two-thirds that of the cylinder. So to find the volume of the sphere, you deduct a third from that of the cylinder. The cone is even smaller – a third of the cylinder – so you need to remove two-thirds of the cylinder to end up with the cone. It follows logically from this that the sphere is twice as big as the cone.

How do we know all this from these three drawings? You can't see at a glance that the volume of the sphere is twice that of the cone. That's why Archimedes was so proud to have proved it that he had these figures put on his tombstone. Today, it is very easy to show that a sphere is twice as big as a cone: we'll be looking at that in the next chapter.

We know this thanks to more recent mathematical

developments, including the number pi (π). Pi is a very special number which can be used to calculate a lot of things, including the area of circles and the volume of spheres. That is very useful if, like Archimedes, you are interested in the volume of round objects. Yet the Greeks didn't know about pi. They had determined that such a number must exist but didn't know exactly how big it was. Here again, it was Archimedes who made the most exciting discovery. Using a calculation that we still don't understand completely, including a figure with ninety-six angles, he concluded that pi must be somewhere between $3^{10}/_{71}$ and $3\frac{1}{7}$, that is between 3.1408 and 3.1428: not a bad result, as it was later calculated at 3.1415... and so on to infinity.

The Greeks didn't get much further than that. Their theories were very clever, but had a lot of limitations: they only used whole numbers and ratios between numbers. Ratios are in fact fractions: $\frac{2}{3}$ is simply a ratio of 2 to 3 but, instead of $\frac{2}{3}$, they wrote them in a more complicated way. They also had no formulas. All of their proofs, including those of Archimedes on volumes, were based on shapes and figures. Luckily, we can now solve problems like this much more simply, but we have the Greeks to thank for the way in which we do that: by using mathematical theorems. Pythagoras, Euclid, Archimedes and many others changed mathematics for good.

The nerds of China

So far, the various ancient cultures seem very similar. Mesopotamia and Egypt started using numbers early on, so early in fact that they were probably among the first things to be written down. There and in Greece, mathematicians enjoyed a high status and largely worked on practical problems, but using general methods.

That was very different in China, and the difference showed itself from the beginning. In China, people probably didn't start writing for administrative reasons, as no long lists of goods and quantities have been found. Fortune-telling, on the other hand, was very important, and the first script was based on marks on the bones used by fortune-tellers. Although the Chinese did start using mathematics at a certain point, it had little status, which is why we don't know much about it in ancient China. What we do know is that relatively late, around 1,000 BC, they made calculations for calendars and for administrative purposes.

They used two number systems to do that. They had words for numbers, as is common in normal spoken language. Those words were – and still are – very simple in form. The word for 354 is spoken as it is written, as 'three hundred five ten four', similar to how numbers are written in English or French, and less complicated than in German or Dutch, where the 4 and the 50 are reversed. The second way of writing numbers was more revolutionary. At first they used bamboo sticks, but these were later replaced by symbols with lines. For the numbers from 1 to 9 the sticks were laid out in a special way and were then repeated for higher numbers, just as numerals repeat the names used for 1 to 9.

The numbers from 1 to 9 in ancient China, written horizontally and vertically

There were even two kinds of symbols for the number system based on sticks. In the illustration, the symbols in the top row are written horizontally and in the lower row they are vertical. The Chinese used these two sorts of symbols to show a zero. In Mesopotamia and Egypt they were unable to distinguish between numbers with and without a zero (for example, 506 and 56). They had no way of writing 0, and certainly not of showing that there were no tens in a number. With their two kinds of symbols, the Chinese were the first people in history who could do that. The illustration shows how they wrote the number 60,390:

The number 60,390 in ancient Chinese notation

They used both sets of symbols together when writing numbers. In the illustration, you can see that the 3 is written using the vertical notation, while the 9 is horizontal. That means that there was no 0 between them. The 6 and the 3 are both vertical, showing that there is a 0 between them. They didn't write the 0, as they didn't yet have a symbol for it. Two vertical (or horizontal) symbols next to each other told them there was a 0 between them. Unfortunately, without a symbol for 0, they were as yet unable to show how many 0s there were between the numbers (in the illustration, the boxes make it clear). Nevertheless, the Chinese system of counting was a great breakthrough because it was the first time that any number could be written down using no more than twenty symbols.

Besides a clever counting system, the Chinese also had

various methods of making calculations. They could multiply quickly in the same way that we still do. To calculate 81 × 81, they first laid out the numbers with sticks and then added to them a step at a time: first 80 × 80, then 80 × 1, etc. They also had ways of solving more difficult problems, which are collected together in a Chinese book on maths called *The Nine Chapters on the Mathematical Art* (*Jiuzhang Suanshu*) and later commentaries on the book. The titles of the chapters give you a good idea of what the Chinese were capable of around the year 0 AD:

1. Rectangular fields (areas of various shapes and fractions)
2. Millet and rice: exchanging goods at different prices
3. Proportional distribution: distribution of goods and money in fixed proportions
4. The lesser breadth: sides of rectangles, circumference of a circle, extraction of square and cube roots
5. Consultations on works: volumes of different shapes
6. Equitable taxation: more difficult word problems on proportional taxation, e.g. in relation to the number of people
7. Excess and deficit: linear equations, e.g. how (nowadays) your income increases when you work more hours
8. The rectangular array: combinations of linear equations, to do with agricultural yields and the sale of animals
9. Base and altitude: applications of what we know as Pythagoras' theorem

For the Chinese, mathematics was not about abstraction. There is not a single general definition or piece of proof to be found in the whole book. Their main concern was to solve practical problems, by giving many concrete examples. They wanted to find the most generally applicable methods possible and, as long as they worked, it was not important to trace them back to basic mathematical principles.

Mathematics was therefore intended above all to be useful. Anyone who learned maths addressed issues of taxation, architecture, warfare and many other things. While in Mesopotamia and Egypt that gave mathematicians a high social status, as managers immediately under the most senior bosses, in China it was very different. There, mathematicians worked with craftsmen to solve problems. They were seen more as the 'nerds' of society. Even when Chinese mathematics was at its peak, mathematicians complained that they were looked down upon by people who had studied literature. A Chinese emperor would never boast about his mathematical knowledge.

And to think that mathematicians played such an important role in China. The most famous book from that period (the *Shushu Jiuzhang* or *Mathematical Treatise in Nine Sections*), written in about 1247 AD, devotes two chapters to fortification works and calculating the distance to the enemy camp, were badly needed because of the war with Mongolia. The book contains a lot of other practical things, such as credit systems and rules on how to build dykes, as well as 'useless' things, problems that are solved in unnecessarily complicated ways. The solution to one problem was so complicated that this thirteenth-century Chinese book contained things that weren't discovered in Europe until 1890!

In short, in China too maths played an essentially practical role, particularly in organisation and administration. But it fulfilled that role differently from the other ancient cultures, with general methods for solving problems rather than abstract proofs, and with concrete examples instead of definitions and basic principles. The methods were different, but the reasons for using maths were the same. This a good moment, then, to return to the question I asked at the end of the previous chapter: why did we start using mathematics?

The answer is actually very simple: maths allows us to organise cities and other large societies. Taxes can be collected without using numbers or maths, but it proves almost impossible in practice. Maths will develop at some point wherever people live and trade in large groups. Planning cities, designing buildings, keeping records of food reserves, producing weapons – we need maths to do all these things. We could do them by relying only on our inborn skills, but we need maths to do them better, more efficiently and more exactly.

There are different ways of looking at that answer. Different cultures had their own systems for writing numbers. Sometimes they are very simple, such as the Egyptian method of writing ½, and sometimes they are overly complicated, like the Egyptians' ⁵⁄₇. And yet all those different approaches, from the abstract one of the Greeks to the example-based one of the Chinese, produced the same result. The Egyptians, for example, could divide up bread effectively and distribute it as pay, and in a way that reflected differences in status. The head of a temple was paid exactly thirty times more than the lowest labourer. That kind of system is a lot easier to organise with numbers than without.

We have already looked at this idea in Chapter 1. There, too, we saw that mathematics makes problems simpler and offers practical solutions. That's why we started using maths in the first place. Cities and countries had administrative problems that were difficult to solve using only our inborn abilities. So we developed maths to help us deal with them. There is a good reason why cultures that don't use maths are small: their people live in villages and know each other. City states and kingdoms are too complex to manage without maths.

And yet we see that the more complex mathematics they developed, all those extra sums, were not really useful. They only showed how good or bad someone was at maths. So is more complex mathematics ever of any use? Are there good reasons to go further than simple numbers and measurements? And do we notice any of that in our daily lives? These are questions we will be looking at in the following chapters.

CHAPTER 5

Change is all around us

I'm driving along a Swedish motorway. For those of you who don't already know, Swedish motorways are incredibly boring: hundreds of kilometres of dead-straight road, lined with trees. Luckily, Swedish drivers are very law-abiding and drive at the same speed, so I can put the cruise control on, sit back and relax. Meanwhile, the car's computer is calculating how fast the car is going, how much that deviates from the speed I want to drive at, and whether it needs to accelerate or slow down. Some luxury cars can even check whether you're driving in the middle of your lane. Their computers look at the distance between the car and the lines on each side of the lane and at the direction you're driving in. If you move too much to one side or the other, the computer will adjust your direction.

It all sounds very clever, but just how difficult is it? We can do all those things ourselves, without all kinds of calculations. If I need to drive at a certain speed, I just look at the speedometer and adjust the gas pedal till I've got it right. And road-users need to go along with the flow of traffic, not stick too rigidly to 120 kilometres an hour. As for keeping in the middle of your lane, anyone can do that, can't they?

Yes, they can. Even a computer. But a computer can't feel how the steering wheel responds or see what the traffic is doing like we can. Computers have to work everything out mathematically, and that's quite an achievement. It wasn't easy to devise methods to enable us to make calculations for changing processes, like the speed of a car or how far you are from the next lane. And yet we have found the maths to do it. That's what your car uses when you switch on the cruise control, and self-driving cars use it even more. Without maths, we wouldn't have these applications.

Isaac Newton is responsible for the mathematical breakthrough that allows us to have cruise control in our cars today. At least, that's what the British say. There was a scientist living in Germany at the same time, Gottfried Wilhelm Leibniz, who came up with exactly the same idea. To explain exactly what that idea was and how important everyone found it even then, we have to go back to the ancient Greeks, to Archimedes and what he discovered about cylinders, spheres and cones.

Archimedes wanted to prove something about volume. You might remember how to calculate the volume of a sphere. There's a standard formula for it that gets rammed into all of us at secondary school: the volume of a sphere is $\frac{4}{3} \pi r^3$. That means we need another two formulas. The volume of a cylinder is the area of a circle times the height of the cylinder, i.e. $\pi r^2 \times 2r$, or $2 \pi r^3$. Lastly, the volume of a cone is $\frac{2}{3} \pi r^3$. It's not important here how we got to these formulas; you don't even need to understand the πr^3. My point is that, once you have the formulas, you can immediately solve Archimedes' problem. How much bigger is the sphere than the cone? Divide $\frac{4}{3}$ by $\frac{2}{3}$ you'll know: the sphere is twice as big. And how much of the cylinder is left after you have made it into a sphere? Divide ⅓ by 2 and you will get the answer: two-thirds. The high point of Greek mathematics is a piece of cake once you know these three formulas.

So why did the Greeks have such a problem with it? First, they didn't have the number π as we use it. More importantly, to find the formulas you have to work with infinity – something the Greeks refused to do. They preferred to stick with whole numbers and fractions, which are clearly finite quantities and have nothing to do with infinity. That is important because you can't write all numbers as a fraction or a whole number: π, for example, can't be expressed as a fraction. Today, we can indicate it with numbers after the decimal point, but the problem is that – in the case of π – those numbers continue into infinity. It starts at 3.1415 and just keeps on going.

The Greeks knew perfectly well that it's not possible to write all numbers as whole numbers or fractions. As we saw in the previous chapter, Pythagoras allegedly threw a pupil overboard from a boat for showing that this wasn't possible for $\sqrt{2}$. They did, however, come up with a completely different answer to this riddle from the one we have today: the Greeks decided that you can't measure everything in the same way. If something is $\sqrt{2}$ centimetres long, you shouldn't measure it in centimetres but choose a different measure in which the same length can be expressed as a whole number or a fraction. In the triangle in the illustration overleaf, for example, where the long side is $\sqrt{2}$ (because, according to Pythagoras' theorem, the length is the square root of $1^2 + 1^2$), the Greeks said that you can't measure all three sides in the same way. You simply use different units (e.g. a measuring stick that's as long as $\sqrt{2}$) and work with all of them.

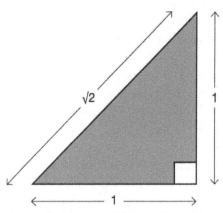

A triangle that caused the Greeks problems

No wonder that the Greeks would never say that the volume of a sphere is $\frac{4}{3}\pi r^3$. That formula is based on π, a number that you can't use to make calculations. The first person to really do that was the Flemish mathematician Simon Stevin (1548–1620). Stevin's work was based on ideas from India and the Middle East that had been adopted by Europeans. Their main step forward was to take fractions increasingly seriously, by writing them as numbers after the decimal point: $\frac{1}{5}$, for example, became 0.2. Stevin summed up the changes that took place at the end of the sixteenth century in one general definition: 'Number is that by which the quantity of each thing is revealed.' He agreed completely that everything can be calculated with a single method of measurement and that we just have to accept numbers like π and $\sqrt{2}$ for what they are.

That was a big step, because it acknowledged infinity: you can, for example, never write π out fully. With fractions, like $\frac{1}{3}$, it is possible, though as a decimal it is written with an infinite number of threes: as 0.3333. . ..

The only difference between ⅓ and π is that, with ⅓, the numbers after the decimal point are recurring and therefore predictable. You know that the next number will always be a 3, while with π you don't know what it will be.

And yet, no one finds it strange if we talk about π. We've become so used to numbers like this that they only seem peculiar when we think about them a little longer. Take the number 0.999... Strange as it may seem, it's easy to show that it's actually equal to 1. As we know, 0.333... is the same as ⅓ as long as there are infinitely many 3s (I hope you'll forgive me for not writing them all down). If you multiply either of them by 3, you get 1. As 0.333... × 3 is also 0.999..., then 0.999... must also be equal to 1.

Infinity soon makes your head spin, but the cruise control on your car wouldn't work without it. Without numbers like π, with infinitely many digits after the decimal point, you can't talk about things that change constantly, like the acceleration of your car. There aren't enough numbers to make the calculations. A car doesn't go from 100 to 101 kph in one step but also passes through 100½ and 100.1415... (with infinitely many digits after the decimal point). Without numbers for all those speeds, they can't be measured in kilometres an hour. Just as you can't measure all the sides of a triangle in centimetres without the number √2.

Newton versus Leibniz

Archimedes didn't have enough numbers to think about volumes the way we can now. As he couldn't use the same units to measure everything, maths was no help either.

Mathematicians were only able to calculate volumes and things that change constantly once they had accepted that there were more than just whole numbers and fractions.

The first to do that were Newton and Leibniz. Between 1660 and 1690, completely independently of each other, they devised a new form of mathematics. Neither could believe that the other had had exactly the same idea. What they had invented has become a famous – and infamous – part of maths known as calculus. Their new method enabled them to measure how quickly something changes and how much it changes over time. These two components of calculus are known respectively as differentiation and integration.

Two mathematicians with their own groundbreaking theory, except that both theories are practically identical. Who was the first? Who should get their name in the history books? That was the big question, especially as Newton was English and Leibniz German. Since the two countries were not on especially good terms with each other at the time, their sensational discovery became a matter of national pride.

It started in 1684 when Leibniz published his discovery, a method to calculate change. Mathematicians were immediately enthusiastic and Leibniz gathered a small group of people around him to work out 'his' new theory in greater detail. In 1693, he even published the first book explaining calculus for a broader public. Newton, on the other hand, published almost nothing. People close to him knew that he had discovered a new mathematical method, but no one knew exactly how it worked. He kept his method as secret as possible so that he would be the only one to use it.

It is not surprising, then, that Newton was annoyed when Leibniz suddenly announced the discovery of the same mathematical method without acknowledging him in any way. Newton had sent Leibniz a letter several years earlier, in 1676, explaining his new method – but in code. That was customary at the time, though the letters weren't always easy to decode. Galileo had told Kepler in a coded letter that he had seen two moons orbiting Jupiter, but Kepler understood him as saying that Mars had two moons.

Newton's letter was intentionally indecipherable. He didn't send it to explain to Leibniz how his method worked, but to be able to say later that the German scientist had stolen 'his' theory. And that's what Newton claimed – or rather he had his pupils claim it. When he saw that Leibniz was distributing his mathematical theory, he ordered his own followers to make the German look ridiculous.

What followed was one of the most unpleasant disputes in the history of science. Even their own contemporaries – who were accustomed to such spats – were shocked. For many years, followers of Newton and Leibniz distributed pamphlets ridiculing the other camp. Leibniz wrote a book defending himself and asked the Royal Society, the most renowned scientific institution of the time, for help. The Society initiated an independent inquiry to determine which of the two scientists had come up with the theory first.

Unfortunately, the inquiry wasn't at all independent. Newton was President of the Royal Society at the time and, although he insisted that the official committee set up to conduct the inquiry would work independently, the committee in reality did nothing at all. Newton secretly wrote the report himself, naturally concluding that he had invented the new method on his own and that Leibniz was a despicable thief who refused to admit his defeat. It only became publicly known 133 years later how far Newton had gone to defend himself.

The report solved nothing, of course. Leibniz persisted in defending his reputation in an 'anonymous' response to the Royal Society's report. Insults continued to be thrown back and forth until long after Newton's death in 1716. Who was right? We now know that Newton had indeed come up with the theory first, discovering integration and differentiation in 1665. At that time, Leibniz was a young man of twenty who as yet knew nothing about mathematics. Nevertheless, he had not stolen Newton's ideas. He simply had the bad luck to come up with the same theory a few years later.

Ever-smaller steps

It was immediately clear that the new mathematical theory would be enormously important – that's why there had been such a fight about it. But what exactly had Newton and Leibniz devised? It was a way of calculating how quickly and how much things change. Before then, it had only been possible to count or measure something that stayed the same. Newton and Leibniz revolutionised that by using infinity and the 'new' numbers.

Calculating the speed of change is useful in all kinds of situations. The cruise control in your car continually has to calculate how much it needs to accelerate or slow down, a self-driving car has to calculate how much to adjust the steering, and your luxury coffee machine calculates how warm the heating element has to be to get the water to exactly the right temperature for an espresso.

It is even used in a hospital to see how fast a tumour is growing.

We use the same technique to do all these things. It's all about measuring change. What kind of change it is isn't important: the maths is the same. That's why it can be explained with a simple example. Imagine you're a police officer and you have to catch people who are driving too fast. That means calculating what speed they are driving at, or how quickly they change position. To start with, you do that with a minimum of maths and modern techniques.

The simplest way is to get a colleague to stand a certain distance further up the road, say a kilometre. You both note when a car passes and then compare the times. The aim is to work out how fast the car was driving when it passed you – i.e. at the beginning of the kilometre – not its average speed over the whole distance. But to find out its speed at the start you need to know how long it takes to cover the whole kilometre. If it takes half a minute to cover the whole kilometre, you can assume that it must have been driving at 120 kph when it passed you.

Or not? Maybe the driver was initially driving at 140 kph and, because the speed limit is 120, he braked when he spotted you. Then he drove the rest of the kilometre more slowly, completing it at only 100 kph. You and your colleague would then come up with an average of 120 kph even though he was driving a lot faster at the beginning.

To stop drivers doing this, you can shorten the distance. It takes fifteen seconds to cover half a kilometre at 120 kph, so anyone driving too fast only has a very short time to adjust their speed. The shorter the distance, the more accurate the speed measured at the start. In practice, at a certain point it makes no difference, as cars can't change their speed significantly in a millisecond. That's why signs that tell you how fast you are driving work quite well. They do this kind of calculation over a very short distance of about a metre.

Imagine that's not enough. You want to know exactly how fast a car is driving at the moment you see it. Then even the small error you make in measuring average speed over a few metres is too big. To make your measurement more accurate, you have to make that distance even smaller. And that's where infinity comes in: if you can make the distance you measure infinitely small, your calculation will be infinitely accurate, and you will know the speed exactly.

Newton and Leibniz were the first to come up with this idea. They thought in terms of how fast a point moves up and down along a line on a graph. The steeper the line, the quicker the point would go up or down.

Look at the curve in the graph opposite. Ignore the two straight lines for a moment. You want to know how quickly the lower point goes up as it moves to the right. So you measure the height of the point at the bottom of the curve and again when it is a little further to the right. You then connect the two points with a line and compare the difference between the two heights. That shows you how fast the point has moved from the bottom of the curve to its position further to the right. The problem is, the measurement isn't correct. The point didn't move up- wards very much at all at the start of the line and increased its speed as it went along, as if the car in our example accelerated.

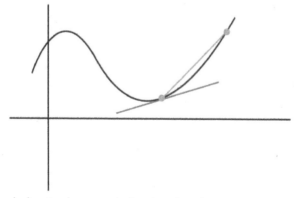

Graph showing how to calculate how fast the curve goes upwards
at the lowest point

Newton and Leibniz solved this problem by moving the second point more and more to the left to decrease the difference between the two points. In this case, the shorter distance made the line less steep and the error less significant. Their idea was to make the distance between the two points infinitely small. The line would then look like the lower line in the graph, which is exactly as steep as the curve at that point. But to do this, you have to calculate with something that is infinitely small.

That was a tough problem for Newton and Leibniz too. In fact, it took hundreds of years for someone to come up with a way of writing it down in a manner that was clearly understandable. After all, isn't infinitely small the same as zero? How can you measure speed over a time of zero seconds? Surely then the car isn't moving at all? The same applies to the lines. Of course you can draw a line between the first two points, but what if the distance between the two points is infinitely small? You can't draw a line between them then, can you?

It's difficult to imagine things like this. That's why it took so long for mathematicians to understand what they were doing. They still did their calculations because they worked, but no one knew how. All because the difference between infinitely small and zero is not easy to imagine, just like the difference between 0.9999. . . and 1.

Eventually, a number of mathematicians came up with the idea of doing away with 'infinity' altogether. It's too difficult, and actually what you're talking about is something that is as small as possible. If you make sure you can always use a smaller distance, you're on the right track. After all, if you discover that you've made an error, you can still make a more accurate measurement. If all this is still too theoretical, don't worry, the details aren't that important. All you need to understand is that the smaller you make your measuring units, the more accurately you can measure the speed of a car.

Why couldn't the ancient Greeks do this? For two reasons. First, they didn't have enough numbers. They might end up with a number like π, which they refused to use. Then the calculations don't work because you need to be able to measure every possible speed. Secondly, they thought the idea of infinity was insane. How can you measure an infinitely small distance? No one can do that. They had a problem with infinity, and perhaps we still do too. After all, even now it's difficult to understand how differentials, which we calculate in this way, work.

Counting steps

Unfortunately, integrals are no easier to understand. Differentials are about speed, about how quickly something changes. Integrals are about quantity, how much something changes. That means counting changes as far as possible cumulatively. If you want to know how big a tumour is after it has been growing for a time, you need an integral. We use integrals whenever we want to know how much something has changed, no matter what it is. It might be the total amount of electricity you've used, the chances of Donald Trump being re-elected, how much a supporting girder can bend, or the damage to your car after an accident. Without being aware of it, we come across integrals everywhere, including in the way car-makers make sure you will survive a collision.

How does that work? Again with small steps, only now we want to follow a large number of small steps and add them all up. Imagine you work as a mathematician in a car factory and have to make cars as safe as possible. You can do that by trying out all kinds of things, wrecking cars and seeing what happens. But you can also do it a lot more cheaply using maths.

When a car crashes, it is mainly the heads of the occupants that are in the greatest danger. The more they are thrown back and forth, the more dangerous the crash is. So the speed is crucial and you can use a differential to measure it. For each moment of the crash, you can look at how fast the head is moving. First it is thrown forwards, where it hopefully hits an airbag that breaks the forward momentum. Then it recoils backwards against the head support, and then forwards again.

So, as a mathematician at the car factory, you first work out how fast the head is moving at a whole series of moments. But you still don't know how dangerous the crash is, only the speed of the head. And that's why you also need an integral. A head moving at high speed is of course dangerous, but moving back and forth for a long time is even more so. Think about it: spinning round once isn't so bad, but do it quickly twenty times in succession and you'll end up very dizzy.

That's a good reason to add up how fast the head moves in total during the crash. If you're lazy you choose just one speed, for example the fastest, and multiply it by the duration of the crash. But that simple sum will suggest that the head moved a lot more than it actually did, making the cars seem less safe than they are in reality.

This is exactly the same problem as with the speed checks we looked at earlier, and the solution is the same: your calculation will be more accurate if you look at a series of much smaller steps altogether. What you really want to do is add up all those infinitely small steps, which will then tell you exactly how much the head moved back and forth, and thus how dangerous the crash was. Car-makers still use this calculation to predict how safe their cars are. Of course they test the cars too, but a mathematical calculation makes it all simpler and more predictable. They don't need to wreck as many cars before finding out how safe they are and they get a safety rating more quickly. Researchers can then identify at what score passengers may suffer concussion, for example. In this way, integrals protect your safety.

Didn't integration also have something to do with areas and volumes? You might remember that from secondary school, as integration is related to the formulas for Archimedes' theorem on spheres, cones and cylinders. They use the same trick, though the change is not so clear; you have to imagine it yourself. The illustrations opposite show how to calculate an area with a whole series of steps that you add up together. The smaller the steps, the better they fit into the space below the wavy line.

It is, however, difficult to see the change in these graphs. It might be easier to imagine them as lying flat on the ground. You want to know the area of the ground below the line. It would have been easier if the area was a rectangle, because then all you would have to do is multiply the length by the breadth. But if you divide up the area below the line into small rectangles, you can use that easy solution. You work out the area of each individual rectangle and add them all up. And the smaller the rectangles, the more accurate the calculation.

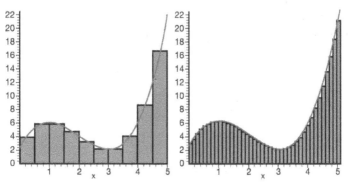

Rough calculation of the area below the line; the smaller the rectangle, the more accurate the calculation

You can also calculate volume in the same way. That's a little more difficult because you have to work both vertically and horizontally, but the principle is the same. Sometimes, all you have to do is use the standard formulas I spoke about at the beginning of this chapter. But it might be a more concrete problem: you can also see calculating how dangerous a crash is in terms of calculating an area. The rectangles below the line refer to the movement of the head. The line shows how quickly the head moves, and the higher the line the faster it moves; the sum of all the rectangles shows you how much the head has moved back and forth in total. The idea is the same; it's just described differently.

Nothing changes more than the weather

At last, the weather forecast for tomorrow is good. But when can you assume that the forecast is right? Weather reports are so often wrong that we should take them with a pinch of salt. At least that's how it used to be, until large computers made it possible for meteorologists to use calculus to predict the weather. Since then the forecasts have been surprisingly accurate compared, for example, to those of the 1970s.

Until then, the weather was predicted using three simple steps. You looked out of the window and studied the clouds, temperature, and so on; then you searched in the records for an earlier day on which the weather was similar; thirdly, you used the weather on the following day back then to forecast what the weather would be the day after in the present. In other words, you assumed that the weather would be exactly the same as it was before, with the coming few days the same as they were then. If you only look at clouds and temperature, that is of course hardly ever right. Predictions in the past were very often wrong because the weather is a little more complicated than that.

Of course, it's also possible to 'calculate' the weather. The changeability of the weather caused by air flows is a perfect application for calculus. In the First World War, for example, the English mathematician Lewis Richardson experimented with using maths to forecast the weather. He started cautiously by trying to forecast the weather for the coming six hours. He would look outside, do a quick calculation and then know what the weather was going to be like six hours later. Richardson was, however, wrong about the 'quick calculation': it took him six weeks!

Forecasting the weather by calculation was therefore very difficult. It not only took far too long, it was often wrong as well. That's because, with the weather, so many things change. The air is constantly moving and temperature, humidity, etc. are consistently changing. You have to know where areas of high and low pressure are and how they are moving – and for a very large section of the atmosphere. Even small changes can make a significant difference.

Because of all those changes, we still can't predict the weather exactly. Even a gigantic super-computer can't calculate quickly enough to do that. So we've given up trying to know everything exactly and have settled for a compromise: a super-computer pretends that the weather is the same all over an area of about ten square kilometres, because smaller areas create too much calculation work. So, although forecasting the weather has become a lot more accurate since we started using this compromise, it is still not 100 per cent reliable.

Should we believe forecasters if they tell us the weather is going to be fine tomorrow? Yes, we should. Although meteorologists can't predict the weather exactly, they can do it very well. Computers work out how the weather is changing between all those squares, using differentials to see how quickly the air is moving and integrals to see how much the weather has changed over a certain period of time. So our weather forecasts have improved considerably, thanks to maths. They are now so good that forecasts for the following day are almost always right. Even forecasts for the following week are right 80 per cent of the time. So those integrals and differentials are pretty useful after all.

Integrals and differentials in buildings, policy plans and physics

The weather isn't the only thing that constantly changes. Though you might not notice it, buildings too are continually affected by things like the wind and the weight of the people walking around inside them. Gravity tugs at them, trying to pull them down to the ground, and yet they remain standing because we have become quite good at knowing how to build solid structures. And that, too, got a little better when we started using maths.

For a long time, buildings were designed on the basis of experience. People built what they knew, without experimenting too much. When they did try new things they were very careful, waiting first to see if they worked. Building was an art until around 1900, when it started to become more of a science. Take, for example, the Golden Gate Bridge in San Francisco. When it was built in the 1930s it was, at nearly three kilometres, by far the longest bridge in the world and was held up by 129,000 kilometres of cable. Everything about it was bigger than anything that had ever been built before. How do you construct something like that? And how do you know that a bridge of that size won't collapse? Or that the wind won't blow it down? Or whether it won't be too heavy in the middle? All that is calculated in advance.

The Golden Gate Bridge in San Francisco*

The physics used to calculate whether a bridge will collapse or not depends on integrals and differentials. They are used mainly to calculate how much steel girders can bend. As you can see in the photograph, the Golden Gate Bridge is largely constructed of girders, which have to bear a considerable weight. That causes them to bend, and it is possible to calculate how far they do so. A differential is used to determine how they change in shape and an integral to find out how far they bend in total. A large number of factors are taken into account in the calculation, including the way in which the girder is positioned. This is explained in the drawing overleaf: a girder that is lying flat bends much more than one that is lying on its side.

* Wikipedia/Rich Niewiroski Jr.

The extent to which a girder changes shape depending on how it is
positioned

Maths takes the guesswork out of building. Builders had
experience with steel girders, but no one had experience with
building a bridge as big as the Golden Gate, or of building
a bridge that big completely of steel. You can just start and
hope that everything will stay the same when you build on
a large scale, but that can get very expensive. Imagine that,
in the end, it goes wrong. Taxpayers don't want to bear
the costs of a series of failed experiments with bridges that
keep collapsing. Fortunately, by doing all the calculations
in advance, that's not necessary. Maths has made it possible to construct increasingly large and complex buildings.
Because we can work out in advance whether a building
will remain standing or not, we can now create buildings
that no one has ever seen before, like the CCTV Tower in
Beijing.

The CCTV Tower in Beijing

Change can be found in many other areas, too. Think about the economy, where money flows from one place to another. The number of jobs, vacancies and people looking to fill them changes constantly. Sometimes governments make plans that will cause changes, and the potential impact of those plans has to be calculated in advance. What effect will different tax rates have or, more widely, Brexit or the trade war between the US and China? Governments get economic research institutes to analyse their plans and work out what that impact might be. To make their calculations, the research institutes use large mathematical models. They have series of formulas that tell them exactly what will happen to the economy if the plans go ahead. And what are in the formulas? Exactly – integrals and differentials.

Tax changes affect how much the government collects from individuals and companies, and therefore the amount of money they have to spend. And that, in turn, will impact on the economy. Governments introduce changes in the hope of improving all kinds of things and the research institutes help them to predict what those effects will be. And that's all a little more exact if you calculate them mathematically. What's more, there's less chance that you miss something. You might forget something, but a formula won't.

We don't hear much about the calculations of the research institutes, except perhaps in the run-up to elections. But integrals and differentials are all around us, also closer to home. In your car, your coffee machine, or the thermostat for your central heating. And in the automatic pilot on your plane when you go on holiday. All of these machines depend on that same piece of maths that all secondary-school pupils complain about so much.

What all these machines have in common is that they have to regulate change. Your thermostat makes calculations to keep your house at the right temperature. If it's 16 degrees in your house in the morning and you want it to be 18 degrees, the thermostat works out how high the heating has to be and for how long. It uses a differential to keep track of how quickly the difference between the actual and selected temperature disappears – in other words, how quickly the house warms up. To prevent the house from becoming too warm so that it has to cool off again, the thermostat calculates differentials and integrals.

The cruise control in your car and the automatic pilot in your plane must make similar calculations. In your car you have to keep your foot on the accelerator, otherwise it will slow down. The cruise control uses differentials and integrals to work out how much it needs to accelerate to keep the speed at a constant level. The autopilot does the same thing, and it is also the idea behind the impressive landings of SpaceX rockets. There are changes to be calculated everywhere, and it's almost impossible to do that without integrals and differentials.

Integrals and differentials are also essential in physics. In the natural world everything changes constantly, so to study it you need a way of understanding change. And that's what calculus does. Newton used it in his theory of gravity. It was all still new back then, so he didn't use it to calculate much. But, as we saw in the previous chapter, his formulations were surprisingly simple and accurate. So simple and accurate, in fact, that the famous twentieth-century physicist Richard Feynman once said that Newton couldn't have written his theory in any other way without making it worse.

Integrals for everyone?

Integrals and differentials are enormously useful. But they're quite a bit more difficult to understand than the arithmetic and geometry discussed in the previous chapter. And will you ever need them yourself in your everyday life? That's something almost all secondary-school pupils ask themselves. The answer depends on what you do because, as I have shown in this chapter, they crop up all over the place. If you design buildings, there's a good chance that you use them. If you're a natural scientist, you'll probably work with integrals and differentials at some point. And the same applies if you design cars or conduct crash tests. Nevertheless, there are still plenty of professions where you'll never come across them.

So the likelihood that you will find yourself having to work with integrals or differentials in your daily life is very small. We can all deal with changes in our lives without having to use calculus. In that sense, unless you choose a profession where you have to use mathematics, you won't need to know about integrals and differentials. And even if you do work with maths, you probably won't have to calculate them yourself. It's getting easier to leave the calculating to a computer.

So is it important to know about calculus for other reasons? Imagine you want to understand figures, if only because the government uses them to work out how much tax you have to pay. If there's an error on your tax demand, you'd want to know about it. But you don't need to know about calculus to check your tax demand. The government uses calculus to make decisions that affect you and, if you really want to understand those decisions – such as the mathematical analyses of government plans – you do need that mathematical knowledge. But generally speaking, unlike the figures used to calculate your taxes, the results of these calculations will not affect you directly.

Yet this doesn't mean that you should complain too loudly about what they teach you at secondary school. The idea behind integrals and differentials might be difficult to understand, but it's not as crazy as it sounds. The mathematical symbols easily distract you from the idea behind it: studying change by dividing it up into the smallest possible pieces. If you want to know how the things around you work, it is essential at least to understand this idea.

Integrals and differentials have changed the world. They've made computers, smartphones, planes and many other modern machines possible and are indispensable in helping us to understand the world better. It's thanks to that understanding that we can work with modern technologies; without it, we would be still be building on what we know from practical experience. That would have made it much more difficult to construct large and diverse buildings and modern technology wouldn't have been feasible. In short, without integrals and differentials we would be living in a very different world.

So they are certainly not useless. We come across them in many more places than we think. We just don't notice them and we don't need to do anything with them ourselves, as we have reached a point where the calculations are done for us. So does everyone need to work with integrals? No. But I do believe that we should all understand the idea behind them, just as we all learn about history. Integrals and differentials provide an essential backdrop to the world around us even though, unfortunately, that backdrop is often presented in a way that scares people off. There's no need to be afraid: the idea behind calculus, and the value of that idea, are simple enough to understand.

CHAPTER 6

Getting a grip on uncertainty

It is autumn 2016. The eyes of the world are on America and the presidential elections and, as usual, we can't wait for the results. We want to know in advance who has the greater chance of winning: Hillary Clinton or Donald Trump.

The predictions have since become notorious. The experts behind the polls claimed that Clinton had between 70 per cent and 99 per cent chance of winning – *99 per cent!* We all know what happened after that: much to everyone's surprise, Trump won. The polls had been way off the mark, or at least that's what it looked like. In any case, the experts who had declared Clinton certain of victory were wrong. How could so many people have been so mistaken? That's the big question, as mistakes like this happen frequently. Take the polls for the Brexit referendum in Britain. They predicted a majority in favour of staying in the EU. Although they were less certain than in the US elections, there was still a clear lead for those who wanted to remain. But there, too, the polls and forecasts got it wrong: a small majority voted against the EU.

So what use are polls to us? If statistics can so easily give us the wrong impression, can we trust them? Yes we can, but not blindly. Because they are so often wrong, it's good to know how they work. After all, polls too are calculations that come from somewhere. And we haven't been making them for so long; there were no polls in ancient Athens, where people also voted on important decisions. They didn't have the maths to predict the results back then, nor in the time of Newton and Leibniz, though they were developing it. It didn't really start until the mid seventeenth century.

Mathematical games

In 1654, the amateur mathematicians Blaise Pascal and Pierre de Fermat became involved in a challenging discussion. Pascal had been challenged by a French nobleman called the Chevalier de Méré to solve a problem. De Méré liked to gamble on games, but sometimes they had to be interrupted without a clear winner. If the king suddenly came for a visit, for example, de Méré couldn't simply carry on with his game. So he wanted to know how the money with which he was gambling should be divided up if a game was interrupted. He asked Pascal, but he didn't know either and started exchanging letters with Fermat. They were trying to find out how to calculate the chances of winning a game or a bet. And that's how the discipline we now know as statistics was born.

Imagine you're playing a game in which you have to win three rounds but must stop when you are ahead by two games to one. How much should your opponent pay you? Two-thirds of the prize money would seem logical, since you've already won two of the three rounds you need to win. But you should actually get more, because it's about the probability of you winning all the money. Pascal and Fermat worked out that the probability is actually three-quarters, so you should be paid that much of the money. How they got that result is shown in the illustration overleaf.

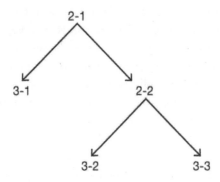

The possible outcomes of the interrupted game

If you won the next round, the score would be 3–1, and you would win. But if your opponent won it would be 2–2 and you would have to play another round so that one of you would win 3–2. That means you win in two of the three possible outcomes. But that poses a problem for the calculation: you don't always play the same number of rounds. If you were to play another round with the score at 3–1, which would end up at 3–2 or 4–1, you would see that you could win in three of the four scenarios. That's why Pascal and Fermat came up with three-quarters.

How useful is this? It doesn't sound like a very important problem to solve. It might come up now and again, but they could just as easily have carried on with the game later. That makes it a surprisingly useless beginning for one of the most widely applied disciplines in mathematics. Remarkably enough, all kinds of mathematicians immediately started making calculations like this for increasingly complex games and for other situations.

Perhaps it wasn't so useless after all: in Fermat and Pascal's time, people were speculating more and more on trade. Investors gambled, for example, that a ship would return safely with a hold full of goods. Sometimes they would change their minds because they needed the money for something else. It's possible that mathematicians devised a simpler version of Fermat and Pascal's method of recovering money that you wanted back before the end of the 'game'.

Whatever the reason, studying games like this didn't immediately lead to anything practical. You needed to know in advance the probability of your winning a round. In my example, both players were assumed to have the same chances of winning each round. But with most games that's not the case. You might be better than your opponent, giving you a greater chance of winning. What you are actually trying to calculate is what will happen if you know everything exactly in advance.

Take the US elections. You can only use Fermat and Pascal's calculation if you know the chances of every registered voter in the country voting for Trump or for Clinton. But that defeats the object; you can't read the thoughts of everyone in America. And if you could, you wouldn't need to make any more predictions, since you would already know the answer. In a sense, you would already have held the elections.

Calculating the probability of something happening is only really useful if you don't already know the outcome. So you start with things you do know, like the answers given by voters in an election poll. Only a small selection of people can take part and you don't know if they'll fill in the answers truthfully, but you have to make do with what you've got. Or you might start with something even simpler, like the colour of a stone. That's how Jacob Bernoulli started, in his 1713 book *Ars Conjectandi*, more than fifty years after Pascal and Fermat. It took that long for someone to realise that it was better to study something of more practical use.

Bernoulli was the first to try to calculate the probability of something happening without knowing all the possible outcomes in advance. Imagine you have a large earthenware pot containing 5,000 black and white stones. You want to know how many are black and how many are white, so you take a few stones out of the pot and see that you have two black ones and three white ones. That might mean that there are 2,000 black and 3,000 white stones in the pot, but you might also have picked out the only three white stones. That's much less likely, but it's possible.

So you keep on taking more stones out of the pot. And there are always two black and three white ones. Logically, you become increasingly certain that there are 3,000 white stones in the pot, just as we are sure that the Sun will rise every day because we have seen it happen so often in the past. But how may stones do you have to take out before you can reasonably say that the ratio of white to black is 3 to 2? That's what Bernoulli wanted to calculate. According to him, you only know it with 'moral certainty' if you get it right 999 times out of 1,000. But then he discovered a problem: he calculated that, just to get it right 49 times out of 50, you needed to take out 25,500 stones!

That's where Bernoulli's book stops. Doing an experiment 25,500 times and not getting remotely close to moral certainty was too much for him. He didn't even publish the book himself – his cousin Johann decided to do so eight years after his death. And that took so long because Bernoulli's widow didn't trust Johann or Jacob's brother, with whom he had argued publicly in scientific journals.

Bernoulli made a good start, but he encountered too many problems. First, you have to guess what the right ratio is. In other words, you have to decide in advance that what you want to know is the probability that there are 3,000 white stones in the pot. The calculation will be different if you want to know the probability of there being 2,999 white stones. Secondly, the number of experiments required and his criteria for certainty were too high. Today, scientists require only that you are right nineteen times out of twenty.

The mathematics of chance started with games and slowly became more practical. Bernoulli was already trying to calculate something more useful. And he got closer to a solution – you don't need to know what all American voters think to make a prediction. But you would need to assume beforehand, for example, that Clinton is going to get 52 per cent of the votes, which wouldn't make it much more practical. After all, we don't know how the whole country will vote. You don't want to take that gamble and fortunately you don't have to, thanks to the ideas of the mathematician Abraham de Moivre. De Moivre figured out our next step by experimenting with something we all associate with chance: tossing a coin.

Tossing coins

De Moivre grew up in France but fled to England in the late 1680s after spending a year in a French prison for being a Protestant. In England he worked as a maths teacher, not in a school but tutoring the children of nobility. In his free time he conducted research and was surprisingly good at it. So good, in fact, that Newton sent people to him with questions on mathematics.

De Moivre also devoted himself to the problem of the black and white stones, though it's easier to imagine it as tossing a coin, something with two possible outcomes, 'heads' or 'tails'. He worked out that if you do that often enough you will get what is known as a 'binomial distribution'. The graph below shows a binomial distribution for tossing a coin ten times. The small block on the extreme right shows the number of times you throw heads all ten times, while the block on the extreme left shows how often you don't throw heads at all. The tall block in the centre shows how often you throw heads five times. This block is the highest, as it is the most likely outcome. In other words, it's more 'normal' than throwing heads ten times (or not at all). Graphs like this crop up everywhere. Height is a good example. The average height of men in the UK in 2017 was around 177 centimetres (5 ft 8in). That means there are fewer men who are shorter or taller. If you are 150cm (5 ft) tall you'll appear on the left of the graph, and if you are 200cm (6 ft 6in), on the right.

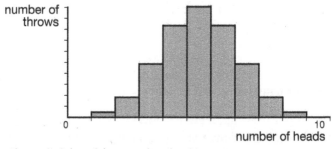

The probability of throwing heads when tossing a coin ten times

The graph showing the results of tossing a coin ten times gives quite a rough picture. The more times you toss the coin, the smoother the curve will become. Look at the second graph, which shows the outcomes for tossing a coin fifty times, and you see that the curve is more gradual.

If you keep going you will eventually get a completely smooth curve, and you can see immediately what that curve has to do with probability. You can use the integrals devised by Newton and Leibniz to calculate the area below the curve. The calculation shows that the top of the curve is 'normal', because almost 40 per cent of the outcomes occur in the two highest blocks.

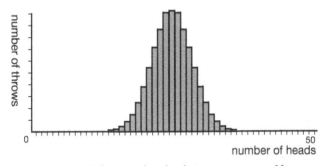

The probability of throwing heads when tossing a coin fifty times

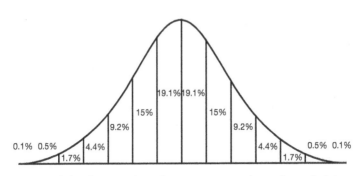

A normal distribution, where the percentages indicate the probability of an outcome falling into each block below the curve

The area below the curve represents a probability: because almost 40 per cent of men are around 177cm tall, there is a 40 per cent probability that any specific man is 177cm. Tossing a coin works in the same way: if you toss the coin 100 times, there is a much greater probability of throwing heads half the time than 100 times. That probability is not zero, but it's very small. And that's why it's so low in the graph.

Two Thomases

De Moivre came up with the graph and the integrals to calculate probability. But what can you really use the graph for? It works fine with body height and IQ scores, but not so easily with more important things like opinion polls, as there are no 'normal' and 'abnormal' votes. Nor does it work so well in science. How can you use a distribution graph, for example, to determine whether you've found a Higgs particle, one of the most important discoveries of the past decade? Is that even possible?

Yes it is, thanks to the work of another mathematician, Thomas Simpson, a contemporary of de Moivre. Simpson elaborated on de Moivre's work and published a book that explained it to a wider public. De Moivre wasn't pleased: in the foreword to a second edition of his own book he addressed himself to 'a certain Person, whom I need not name, out of Compassion to the Public, [who] will publish a Second Edition of his Book on the same Subject, which he will afford at a very moderate Price, not regarding whether he mutilates my Propositions . . .' Simpson responded with equal rancour, but fortunately friends of de Moivre intervened before the argument got out of hand.

Simpson did come up with a new idea: he turned the probability calculation on its head, focusing not on how often you get something right, but how often you get it wrong. In other words, he looked at the probability of the outcome of a scientific experiment being incorrect. Most of the time your scientific apparatus works perfectly and you only make small errors of measurement; you then come out in the centre of the graph, at the top of the curve. But sometimes you will be way off track and make a serious error of measurement. It won't happen very often, because it requires a lot of bad luck. As the probability of making serious errors is small, you end up on the extreme left or right in the graph, at the bottom of the curve.

If everything else goes as it should, thanks to maths we can calculate the probability that our expectations – for example, that the Higgs particle exists – are correct. After all, we don't know which measurements are incorrect and therefore whether we have found a Higgs particle or not. We also don't know whether the measurements that suggest we have found one are correct; perhaps they are the ones that are incorrect. So we assume that our conclusions are incorrect and use the graph to see how strange the measurements would be if that were true. In other words, we calculate the probability of seeing the measured results if there were no Higgs particle. If we need to make a lot of errors in measurement to achieve our results and they end up at the bottom of the curve, that's good news. It means that it's improbable that there is no Higgs particle, and therefore that there's a good chance that it exists. If, however, we need to make hardly any errors in measurement to explain what we have measured, and the results come out at the top of the curve, it means that the Higgs particle probably doesn't exist and we have to disappoint the scientists. Fortunately that didn't happen to researchers at CERN, the European Laboratory for Particle Physics in Geneva. They obtained results that would have been very coincidental if the Higgs particle didn't exist: the probability of getting the same results purely through error was minuscule: 1 in 3.5 million!

Simpson didn't, of course, come up with all this himself. Let's go back to the two problems with Bernoulli's work: it needed too many experiments and you could only calculate the probability that your guess was right. Simpson solved the first problem, because he showed – through more accurate calculations – that you need far fewer experiments to obtain Bernoulli's high degree of certainty. Later in the eighteenth century, another Thomas – Thomas Bayes – solved the second problem by elaborating on Simpson's idea. Thanks to Bayes, we can now also calculate how strange it would be to get the results described above if there were no Higgs particle.

Some degrees of probability are easier to calculate than others. Imagine you get an email and your provider has to calculate the probability that it is spam. One way is to look at how often certain words – such as 'Nigerian' and 'prince' – occur. That is in itself not very difficult, but emails containing these words may not necessarily be spam. So you need to know the probability that an email with those words actually is spam, and that is rather tricky to figure out without any context. Thankfully Bayes devised a formula that allows us to solve exactly that problem:

$$\frac{\text{Probability of spam x probability of words in spam messages}}{\text{Probability of spam with certain words}} = \text{Probability of words}$$

To use the formula, your email provider therefore needs to know three other probabilities. Fortunately, they are a lot easier to calculate than the probability that messages containing certain words are spam: your provider can find them directly in what you put in your spam box. The first thing your provider needs to know is how often you get spam emails, in other words the probability that an email message is spam. To do that the provider divides the number of emails in your spam box by the total number of emails you receive. The second probability is the chances of an email containing the words 'Nigerian prince'. Your provider calculates that by adding up all those emails and dividing them by your total number of emails. Lastly, there is the probability of you receiving spam emails containing 'Nigerian prince'. That, too, is a simple sum: your provider divides the number of emails in your spam box containing the words 'Nigerian prince' by the total number of spam emails you have received. So each calculation you need to make to predict whether an email containing the words 'Nigerian prince' is spam is very easy. As long as the words mainly occur in spam messages and you are not really exchanging emails with princes in Nigeria, you can assume that all these emails belong in the spam box.

We use Bayes's formula quite a lot, because it offers a good solution to Bernoulli's problem. Bayes could calculate probability without having to take a gamble. Of course, the formula isn't perfect, because you don't know whether the probabilities you use on the right side of the equation are correct. They are often easier to check, but there is always a certain degree of uncertainty. Bayes's formula differs from previous probability calculations in that it is of some use in practice.

At the doctor's, for example: if you take a test to see whether you have cancer, you want to know what it means if the test shows that you have the disease. How reliable is the test? What is the probability that you actually have cancer if the test is positive? That can be calculated on the basis of three other probabilities, using Bayes's formula. First, the number of people with cancer: let's say that is 20 in 1,000, or 2 per cent. Secondly, you want to know the probability of the test being positive if you are suffering from cancer, i.e. the chance that the test can detect cancer among people who have the disease: that is in 90 per cent of cases, or 18 of the 20. Thirdly, you want to know the chances that the test shows positive and you don't have cancer. Let's say that is 8 per cent, or 78 of the 980 people without cancer. That tells you that the total probability that the test says 'yes' (for people with and without cancer) is 18 + 78 = 96 people in 1,000. In Bayes's formula these three probabilities look like this:

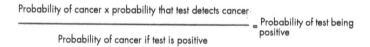

$$\frac{\text{Probability of cancer x probability that test detects cancer}}{\text{Probability of cancer if test is positive}} = \text{Probability of test being positive}$$

If you put those figures into Bayes's formula, you will get the following result: 0.02 (i.e. 2 per cent) x 0.9 (the 90 per cent) / 0.096 (our 9.6 per cent, or 96 in 1,000) = 0.1875, or 18.75 per cent. This shows that the probability of you really having cancer after the test proves positive is only 18.75 per cent. That is much lower than you would expect in a test that gives the right result in 90 per cent of cases of people who actually suffer from the disease. This happens because the test says 'yes' to a lot of people without cancer: as we have seen, of the 96 with a positive test result only 18 actually have cancer. So it is good that we have this mathematical formula, as it allows us to find out how much a test like this actually tells us.

All a game? Maths in practice.

Statistics, the applied version of the probability theory we've been looking at so far, started a little later with a practical problem. The astronomer Tobias Mayer came up with the solution in 1750. And it wasn't an abstract theory, but a piece of maths straight out of the real world.

In Mayer's time, the major European powers had a serious problem. They all had overseas colonies and their ships ploughed back and forth on the world's oceans, but no one knew how to calculate exactly where their ships were; and lost ships cost a lot of money. The British offered rich rewards to anyone who devised a way of calculating latitude and longitude at sea. From 1730 they could determine latitude with a sextant, but longitude remained a sticky problem. So the state sponsored research to come up with a solution. Between 1714 and 1814 it awarded £100,000 – millions in today's money – to people with ideas on how to calculate longitude while at sea. In 1765, three years after his death, Mayer won £3,000, the equivalent of almost half a million today. He had found a way to predict the position of the Moon, and if you know that you can work out what time it is in London. You can then calculate your longitude on the basis of time zones. The baseline is in Greenwich, in London. The further east you go, the earlier it is. In New York it is five hours earlier and in Amsterdam it is one hour later. Once you know the time in London, based on the position of the Moon and the time where you currently are, based on where the Sun is at its highest point (local noon), you can calculate how far east or west of London you are.

The position of the Moon was usually calculated using three measurements, but Mayer used no less than twenty-seven. That was an exceptionally high number for the time, though very few by today's standards. We have become used to large quantities of data, but before Mayer people simply didn't know how to deal with all that extra information. To determine the position of the Moon they needed to know three things, and that meant making three measurements – no more, no less.

That's what Leonhard Euler (1707–83), one of the most talented mathematicians ever, thought too. To make it easier to see why this problem is so difficult, imagine trying to draw a straight line on a graph without knowing where it starts or how steep it is. Without these values, you can't draw the line. If you have a single point on the graph, you know how high the line starts, but not how steep it has to be. You can see this in the graph on the left. If you have two points, as in the middle graph, it's a lot easier: you can simply draw line between them. But what if you have more than two points? The graph on the right has three points. How can you draw the line now? If you draw it between two points, as in the middle graph, you don't use the third point at all. So should you draw it somewhere between the points? If so, where? How steep should it be? And should it start somewhere just above the lowest point? As you can see, it's not so easy to draw the best line between more than two points. And that's why Euler didn't come up with a solution to the problem of how to determine the position of the Moon using only three measurements.

Drawing a line between one, two and three points

Mayer solved this problem in a very simple way: he had three unknowns, so he divided his twenty-seven measurements into three groups of nine. He then took the average of the nine measurements and used the average of those groups as if they were the actual measurements. So he used all twenty-seven measurements to make the three calculations. And it worked: he was able to determine the position of the Moon much more accurately than his contemporaries.

Euler thought Mayer's solution was nonsense. More measurements increased the risk of making mistakes. If you constantly came out two degrees too high, your error would be compounded as you made more measurements. That's why Euler thought it best to use as few data as possible. We know now that he was wrong, but why? Let's go back to the curve on the probability graph. There can be errors anywhere, on the left or the right. Euler thought that the more measurements you added together, the more you would slide down the curve. But because the errors occur on both sides of the curve, they cancel each other out. If you add up the positive and negative errors you end up in the middle, at the top of the curve. And because errors in measurement are random, the more measurements you make the better.

More data!

Around 1800, Mayer's practical work and the theoretical work on probability came together. This was partly due to the efforts of scholars like Carl Friedrich Gauss, Pierre Simon Laplace and Adrien-Marie Legendre, resulting once again in squabbles about who came up with the ideas first. Gauss even got his friends to testify that they had heard him talk about it before any of the others had put a single letter to paper. Who was first doesn't really matter, but it is clear that they all thought they had come up with a very important discovery. That's not so surprising; even before Laplace's death in 1827, dozens of books had been published elaborating on their work. Their mathematical methods were immediately taken up within science, and increasingly elsewhere. A century and a half after Pascal and Fermat, the discipline of statistics took a great leap forwards.

The reason was an improvement in Mayer's method, which was really nothing more than a trick to get round the problem. Mayer didn't change the original calculation, but simply took the average of the three groups. Gauss and Laplace solved the problem a little more effectively. They devised a calculation that shows how to draw a line when you have more than two points. In the graph, you can see a series of measurements that can't be joined by a straight line.

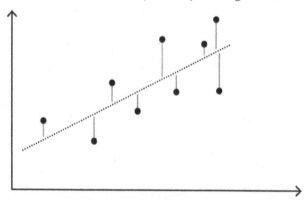

Finding the best line between multiple points using Gauss and Laplace's least square method

Gauss and Laplace showed that the best line between the points is the one that reduces errors in measurement (i.e. results that do not fit perfectly on the line) to a minimum. The errors are shown by the vertical lines between the points and the dotted line. Because they are both positive (above the line) and negative (below the line), adding them all together can produce zero. To calculate the total error, you therefore use the square to cancel out the minuses (the square of -2, for example, is 4). That's all there is to it. By focusing on errors, like Simpson, you can make better use of multiple measurements than Mayer did. That makes your predictions better: if you use nine times as many measurements, like Mayer, your predictions will be three times as good. Mayer's achievement may not have been a massive step forwards in terms of accuracy, but it did make enough of a difference to win half a million pounds.

What's more, Gauss and Laplace's method enabled us to know how good an estimate is, because we know the errors in our measurements. A measurement is more accurate with a lot of small errors than with a few major ones. That was really new. Compare it, for example, with the estimates the ancient Mesopotamians used to make. They guessed at the quantity of grain an area of land would produce, a fixed amount per square metre. But in practice that was, of course, inaccurate as not all land is equally fertile, rainfall is not the same everywhere, and not every farmer takes as much care of their crops. They knew that in Mesopotamia too, of course, but they couldn't do much about it because they didn't have sufficient knowledge of maths to work out the best estimate, or even how good their estimate was. We couldn't do that until Gauss and Laplace discovered how to calculate a best estimate.

What John Snow did know

It was another 100 years before statistics started to be used everywhere, for example to study the causes of disease.

Around 1850, cholera was a serious problem, mainly because no one knew how it spread, leading to regular epidemics. There were a number of theories about it: a lot of people thought it was caused by breathing in bad air or smells. Even more bizarre was the idea that getting angry increased your chances of catching the disease. In 1832 and 1844, the people of New York were advised to stay cheerful and calm to stop themselves from succumbing to cholera. Fortunately, others had the right idea: cholera spread through water, whether people were angry or not. There was, however, little systematic research into the causes of the disease; the whole discussion was purely theoretical.

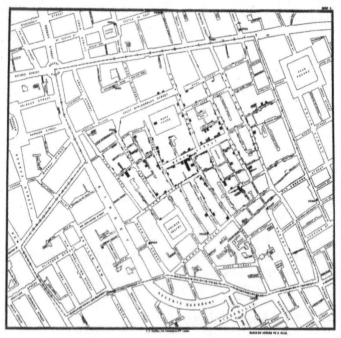

The cholera epidemic around Broad Street, London. The black blocks show the victims.

And then, around 1850, the British doctor John Snow started to investigate the disease. At that time there had been a number of cholera epidemics in quick succession. After his first study in 1848, Snow was able to identify where the disease had come from: a seaman called John Harold was the first to contract it. But that didn't explain why the next occupant of Harold's room also became ill. That called for further research.

Fortunately – at least, for Snow – there was a second outbreak of cholera a few years later. This time he was better prepared. He drew a detailed map of all the locations in London where someone had died of the disease, marking each patient who died with a black block.

Snow discovered that they were all located in the same part of London, around Broad Street. He correctly surmised that the water pump in Broad Street was infected with cholera, as everyone who used water from it got sick. Only the local brewery and poorhouse, which had their own pumps, were spared.

The best anecdotal proof that the disease spreads through water was the story of an old woman from a completely different part of the city who also caught cholera. She used to live in Broad Street and had water from the pump there brought daily because she liked it much better than the water from her own pump.

Real scientific research, however, needs to be more rigorous, and Snow did precisely that some years later in 1885, when a much worse epidemic broke out and claimed thousands of lives. Although he wasn't aware of it, he conducted one of the first double-blind experiments in history. This is a study in which neither researchers nor patients know which group the latter are in (in this case, the clean or polluted water group). Snow thought that, if the disease was caused by water, there might be a link between the company that supplied it and the probability of contracting the disease. So he focused on the two largest water companies in London, the Southwark & Vauxhall and the Lambeth, and discovered that the former drew its water from a more polluted part of the River Thames than the latter. As could be expected, people who got their water from the Southwark & Vauxhall were at greater risk of dying from cholera. The company supplied water to 40,000 households, from which 1,263 people died of the disease. Snow calculated the number of deaths per 10,000 households at 315. The Lambeth company supplied much cleaner water, with the result that 'only' 37 people died per 1,000 households. A smaller company, the Chelsea, used the same polluted water as the Southwark & Vauxhall but filtered the water more carefully, so that fewer of its customers contracted cholera.

Snow was convinced. All his research suggested that cholera spread through polluted water. He was right and, not long after, cholera bacteria were discovered. The only problem was that Snow couldn't show how probable it was that he was right – in other words, that there was a very close link between the number of cholera-related deaths and the water company. By no means all of his contemporaries believed that his experiments proved that the disease was caused by polluted water. There were still doctors in 1892 who believed that cholera spread through the ground. With a little mathematics, Snow could have shown the probability of his being right. Not having that maths at his disposal literally cost lives.

Nicolas Cage and swimming pools

What was missing? How could John Snow have calculated how close the link was between the water suppliers and the number of deaths? We've already seen one way: as with the Higgs particle, you could determine how probable it is that there would be such a difference in the number of deaths (315 and 37 per 1,000 households) if cholera was not spread through polluted water. Can such a large difference be a coincidence? Where would you end up on the probability curve if you assumed that cholera was caused by something completely different? Exactly, somewhere at the bottom. If the probability of that coincidence is very small, then you can safely say that the difference in the quality of the water was responsible for the difference in the number of deaths.

There is also a second way. Imagine there was a series of epidemics in which the number of people supplied with polluted water varied. The newspapers said that water supplied by the Southwark & Vauxhall was dangerous and users switched en masse to water from the Lambeth company. You could then see whether that affected the number of people contracting cholera. If more people drank clean water and there were fewer deaths, you could use those figures to make a calculation.

Such a link – in this case between the number of people who drink polluted water and the number that contract cholera – is called a correlation. As scientists are fond of pointing out, a correlation is not the same as a causal link. A correlation between polluted water and the incidence of cholera doesn't mean that the one causes the other. With a little imagination, you can find correlations between a lot of things. Like, for example, the number of films starring Nicolas Cage and people who drown in a swimming pool.

Look at the graph and you can see that, for many years, there is an eerily close correlation between the number of deaths from drowning in a swimming pool and films starring Nicolas Cage. Does that mean that the actor causes people to fall into swimming pools? Of course not. But there is a correlation, and we want to be able to calculate just how close it is.

Since 1900 we have been able to do that. To work out how close the link is between variables like Nicolas Cage and swimming-pool drownings we use a correlation coefficient, a number between -1 and +1. A negative coefficient of -1 means that, every time a new film appears starring Nicolas Cage, fewer people drown in swimming pools. The two lines are a complete mirror image of each other. A positive coefficient of +1 means the opposite: the more films Nicolas Cage appears in, the more people drown. The number of people drowning in pools doesn't go up unless the number of Nicolas Cage films goes up. In other words, the two lines correspond exactly. A score of 0 means that the two things have nothing at all to do with each other.

Even a correlation coefficient doesn't make it possible to remove all nonsensical correlations. The coefficient in the graph is still 0.666, a reasonably strong positive correlation. That is not so surprising after all, as neither trend changes very much. Nicolas Cage can't appear in twenty films simultaneously and, fortunately, drowning in a swimming pool is nearly always an accident and people don't suddenly have ten times more bad luck than usual in one year. If you search long enough you can always find something that doesn't change very much.

That's why we have to be careful with correlations. In this

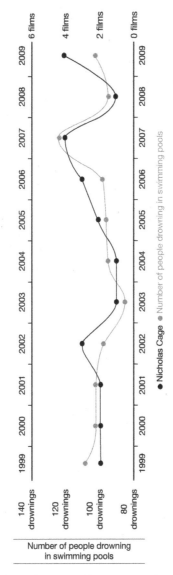

The number of Nicolas Cage films compared to the number of
people drowning in swimming pools

example, it's obvious that Nicolas Cage isn't really responsible for people drowning in swimming pools. But that's by no means always the case. According to an article in *The Wall Street Journal*, there is a correlation between safety in playgrounds and child obesity. Does that mean we should now take our children to more dangerous playgrounds? Does being safe make you gain weight? Most likely not, but someone noticed that playgrounds are getting safer and children are suffering more from obesity and came up with a strong positive correlation – one that the newspapers picked up on. Statistics can easily be misleading.

Fake news? Distorting the world with statistics.

It is quite easy to use statistics to present a distorted view of the world. That has been going on since statistics has been around: a book called *How to Lie with Statistics* by Darrell Huff was published as long ago as 1954. It describes not only strange correlations, but also many other ways in which figures can be misleading.

More recently, in a speech in mid 2017, the then US Attorney General Jeff Sessions said that America was becoming an increasingly dangerous place to live. The number of murders had risen by 10 per cent compared to the previous year, an increase that had not been seen since 1968. Sessions suggested that foreigners were to blame, that the time had come to suspect everyone who entered the country and that immigrants who already lived there were not to be trusted either.

Sounds convincing, doesn't it? But, despite the statistics, America is now much safer than it ever has been. The increase is only so high because the total number of murders is very low: 10 per cent can mean one more in 10, or 1,000 more in 10,000. The number of murders in the US had fallen to such a low level that a small increase immediately seemed very large when Sessions expressed it in terms of percentage.

Sessions's 10 per cent also concealed something else. Almost a quarter of the increase in murders was due to the fact that many more people had been killed in Chicago (765 out of 17,250 nationwide, whereas the largest city in America, New York, had had only 334 murders). The rest of America had largely become safer than it had ever been. The numbers were right – Sessions wasn't simply lying – but the implications were wrong. A cleverly selected statistic presented a completely distorted view of reality.

It's possible to do that in all kinds of ways. If you're interested in whether we're better off than we used to be, you might want to know if we have more money to spend. America has statistics for that. Two, in fact. According to the official figures, collected by the US Census Bureau, average income in America has risen hardly at all since 1979; it even fell for a long time. So things may not have been better in the past, but they certainly weren't worse either.

The other statistic comes from a think tank, not from the government: people have one and a half times as much money to spend as in 1979, so everyone is quite a bit better off. Americans have never had so much money. In fact, incomes have almost always been rising. The government and the think tank paint completely different pictures. So who's right?

Most probably the think tank. The Census Bureau forgot something very simple: they used the figures for average income per household and divided them by the number of people in the average household. But they divided the average income for 2014 by the same number of people per household as in 1979. In the meantime, however, households have got smaller: more people live alone or have no children, and households with children often have less money to spend. So the income for 2014 should have been divided by a smaller number of people. It's logical that things don't seem to be getting better if income is divided among an increasing number of people, in relative terms.

Sometimes statistics are simply too difficult to understand. Take the difference in salaries between men and women. In rich countries, on average, women earn 85 per cent of what men earn. That sounds clear and completely unacceptable. Of course it is a problem, but perhaps not as bad as that single statistic might suggest. Women in those countries are not badly underpaid compared to men in the same job: they actually earn 98 per cent of what men earn doing the same job at the same company. It's still ridiculous that there's a difference, but it's not as significant as the first figure suggests.

The discrepancies in salary are not caused by differences in pay for the same work. The figure is based on the average pay of all men and all women. Women get paid less on average because there are fewer women in highly paid jobs. There are fewer women at senior-management level in big companies and more women in certain professions, such as nursing, which tend to be less well paid than occupations where men predominate, such as the police. So there are clearly serious problems, but they are different from what you might suspect if you only look at the figures for average salaries. Women should have better access to top jobs – through better arrangements for pregnancy and maternity leave, for example – and work done mainly by women should be better paid, but being paid less for doing exactly the same work fortunately doesn't happen very often.

Statistics can distort the world so easily because they often work with averages. The income-rise figure is an average, based on dividing the number of households by the number of people in an average household. And the statistic on the pay difference between men and women is an average, too. Averages don't always give a clear picture of what's happening behind the scenes. It's not immediately obvious that the difference is caused by the fact that men have different jobs to women. Look at the four graphs below. The measurements are in completely different places, and yet the resulting statistical data are the same. The Gauss and Laplace method produces the same line for the best prediction on all four graphs.

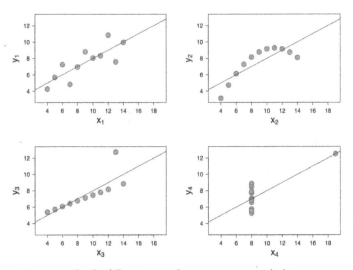

Four completely different sets of measurements with the same predicted outcome

That's why you have to be careful when reading statistics.

It's almost always possible to find a statistic that confirms your world view. If you think that everything was much better in the past, you probably won't believe that we now earn one and a half times as much as we used to. But luckily, there's an official statistic that agrees with you. Or perhaps you think that immigrants make your country unsafe? Then a 10 per cent rise in murders is grist to your mill. The same applies the other way round, of course. Anyone who thinks all the talk of salary differences between men and women is rubbish can quote the statistics showing that women are paid almost the same as men for doing the same job at the same company. And they are right, but that's no reason not to do something about the inequalities that do exist.

Despite these risks, averages can be very useful. They help us understand complex situations quickly. How else would you ever obtain a clear idea of the pay of men and women in rich countries? It's far too much work to compare all their salaries one by one. We need averages to acquire an overview of all that data, just as we need a method of calculating probability and making predictions. Those predictions, whether they are used for the harvest, to determine positions with GPS or make photographs sharper, have become better and more practical through the use of mathematics. And they are used, too, in the opinion polls we talked about at the start of this chapter.

No need to ask everyone individually

Election polls have been around for quite a while. For a little over a century now, we've had the maths to predict how the whole population will vote without having to ask everyone individually. It's actually quite a simple idea. Imagine you want to know what percentage of people think that Donald

Trump is doing a good job. That might be, for example, 40 per cent of the population. But to find that out, you don't ask everyone what they think of him – that's far too much work. The idea behind a poll is that you can ask a smaller group, as long as they are chosen at random. If everyone has an equal chance of being in that smaller group, there's still a 40 per cent chance that someone in the group will think that Trump is doing a good job. In other words, that smaller group is a good reflection of the rest of the country.

Maths is used in polls mainly to calculate how reliable they are. You might choose the sample to be polled at random and still end up with only Trump fans. The more people you poll, the lower the probability of that happening, and that will make the results more accurate. That is if everything goes as it should, since it's very difficult to really select people at random. Take the 1936 US presidential elections. America was in the final phase of the Great Depression and there were serious economic decisions to be made. Everyone wanted to know which of the two candidates would win, the Democrat Franklin D. Roosevelt or the Republican Alf Landon. *The Literary Digest*, an influential weekly magazine, decided to organise a poll among its 10 million subscribers. That was nearly 10 per cent of the population, around 125 million in 1936. Of that 10 million, some 2.4 million eventually took part in the poll.

The magazine published the results of the colossal poll shortly afterwards. They predicted that Landon would win, with 57.1 per cent of the votes. But when the elections were held, the *Digest*'s poll proved completely wrong: Roosevelt won with a landslide victory of 60.8 per cent, while Landon only received 36.5 per cent of the vote. What went wrong? Despite the enormous scale of the poll, the selection was not genuinely random. The *Digest* had based its sample on telephone directories, its own subscribers and membership lists of clubs and associations. In the Great Depression only the wealthy could afford a telephone, a magazine subscription and club membership fees. As they were also more likely to vote Republican, the *Digest* had mostly polled people who voted for Landon.

We haven't seen a fiasco on such a scale in recent times. The polls for the 2016 US elections were completely wrong, with experts predicting that Hilary Clinton had between 70 and 99 per cent chance of winning. Yet, strange as it may sound, the 2016 polls were among the most accurate since 1936; they weren't actually as wide of the mark as they seemed. The chances of Clinton winning aside, the polls gave Clinton 46.8 per cent of the votes and Trump 43.6 per cent. It is mainly the difference that is important, only three percentage points. In the end, Clinton received 48.2 per cent of the votes and Trump 46.1 per cent. The difference in actual votes was slightly smaller than predicted, at 2.1 percentage points. Nevertheless, the polls had correctly forecast that Clinton would get more votes than Trump. It was due to the nature of the US voting system that Trump nevertheless made it to the White House.

All in all, three things went wrong. The selection was still not completely random. Polls have gradually improved over time, but people with a university education respond to them more often than those without. As they were more likely to vote for Clinton, the polls missed out on a significant proportion of Trump voters. Just as in 1936, it remains difficult to include the poor and low-educated in polls.

Secondly, it wasn't easy to conduct a reliable poll in the states that gave Trump his victory. According to the polls, Pennsylvania, Wisconsin and Florida would vote for Clinton; that was how they had voted in the past. But in 2016, a lot of people in these three states didn't know right up to the week before the elections which way they would vote, and almost all of those in doubt eventually voted for Trump. No poll could have predicted that – the voters themselves didn't know at the time the polls were conducted.

Thirdly, there were people who simply didn't say that they were intending to vote for Trump. Whether they really hadn't decided or were ashamed to admit it, we don't know. The fact remains that the polling bureaus more often got a clearer answer from Clinton voters. That, too, wasn't the fault of the bureaus. You can't force people to tell the truth when they fill in questionnaires. The only genuine error in the polls was the imbalance in the educational level of the sample. The other factors only became clear later. All they really got wrong was not foreseeing the turnaround in Pennsylvania, Wisconsin and Florida. The rest they got right.

These examples show that statistics by no means always give a perfect picture of the world around us. Polls can often be wrong, even if they are conducted accurately. Averages can be misleading and there can be close correlations between things that have nothing to do with each other. That's why it's useful to have some understanding of statistics: to know how an average is arrived at or realise that a correlation says nothing more than that two graphs look alike. Statistics can mislead us, but they can also be very useful.

We have already seen that statistics can be used to calculate the probability that you actually have cancer if a test shows that you do. And that the probability can be a lot lower than it seems to be if you don't calculate it. In that way, statistics can give you a greater grip on your uncertainty. Other figures, such as averages, offer a quick overview of a mass of information. They summarise it for you, but don't give you a perfect picture of the situation. We don't have time for that. We can't, for example, read all the information available on the economy. Then a few averages that give us an idea of how better or worse things are can be much more useful.

So is it important to understand this area of mathematics? As with calculus, you probably won't need to use it yourself in your daily life. But it is useful to know something about it. After all, we get a lot of our information from polls and statistics. And they can mislead us in all kinds of ways. Attorney General Sessions gave his fellow Americans a distorted picture of security in their country by the clever use of statistics. Polls can be wrong, by accident or intentionally, due to the way in which their samples are selected. Practically all scientific research makes use of statistics to determine whether the outcomes of experiments could just be coincidental.

In this way, statistics have a great impact on our lives. What is best for your children? How can you stay healthy? What will be the outcome of the next elections? What causes cholera? And: how does a computer understand what is on an image? How does your email provider know which emails are spam? As soon as we have to deal with large quantities of data, we use statistics to make sense of them. It is by far the best way to analyse all the information. And that's why statistics has that great – and expanding – influence on our lives. Behind every piece of data you read there will be a statistical calculation. Think how often you see or hear averages or percentages in the media. Statistics aren't facts that you should just accept unquestioningly;

they come from somewhere. If you understand how that information is calculated and how it can go wrong, you can look at it with a more critical eye.

CHAPTER 7

Mind puzzles

In the early eighteenth century there was a puzzle going around about Königsberg, now the Russian city of Kaliningrad. A river runs through it in which there are two large islands. There were seven bridges across the river connecting the islands to each other and to the mainland. The map shows Königsberg as it was at the time, with circles marking the bridges. The puzzle was: is it possible to walk through the city, crossing each bridge only once?

Map of Königsberg, showing the famous seven bridges

One way of solving the puzzle is to try out a large number of different routes through the city. But that would take you a very long time, since Leonhard Euler – whom we already encountered in the previous chapter – showed in 1736 that it was impossible. This was only one of Euler's achievements: he also devised many new concepts in mathematics, including sines, cosines and tangents. Even when he gradually lost his sight he continued to think about maths, and said that blindness helped him to focus his mind because there was less distraction.

Euler thought it would be easier to work out the puzzle by ignoring all irrelevant information. The map of Königsberg, for example, has nothing to do with the problem, only the bridges. So he drew the bridges as lines, with the islands and mainland as circles. You can only walk from one bridge to another if they are both connected to the same circle. Euler's drawing of the Königsberg puzzle is an example of what is now known as a graph. Confusingly enough, it's not the kind of graph you probably learned about at school, with axes and a line. In this chapter graphs mean the kind of diagram you see here, which mathematicians use to study networks.

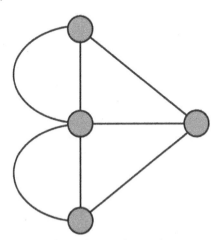

The seven bridges of Königsberg, shown as a graph

It doesn't matter how you walk through the graph – in a circuit starting and ending at the same place, or starting in one place and ending in another. The important thing is that you cross every bridge only once. If you make a circuit, there must be at least two lines going to your starting and finishing point, because you're not allowed to use the same line twice. In the second scenario, with a different starting and finishing point, there must be one line leaving the starting point and one line arriving at the finishing point.

Between those two points, you walk from circle to circle, arriving across one bridge and leaving again across another. At each intermediate stop you therefore walk along two lines. You can't cross two bridges at the same time, or cross by boat to avoid a bridge you've already crossed.

If you take all those factors into account, you will see that it is only possible to walk across all the different bridges in two cases. If you make a circuit, there must be an even number of lines touching each circle in the graph: two (or four, or six, depending on whether you visit the same circle several times via different bridges) at each intermediate stop and two at the starting and finishing point. If you walk from A to B, you need two circles with an uneven number of lines: the beginning and the end. All the circles en route still need an even number of lines, but at your starting and finishing points you need a single line by which to leave and arrive, so there the number of lines will be uneven.

It doesn't matter if it's hard to picture all this. The point is that Euler concluded that it would only be possible to walk across all the bridges once if there were no more than two circles with an uneven number of lines. As Königsberg has four circles with an uneven number of bridges, it is impossible to walk through the city and cross each bridge only once. It doesn't matter how hard you try, you'll never be able to do it.

You might think that all this is not much use in your everyday life. But, just as games led to the development of probability theory, graph theory started with this puzzle. Euler was the first to come up with the idea of solving it by describing it more abstractly with circles and lines. And that idea is now used to plan routes with Google Maps.

One-way traffic

In Königsberg, it didn't matter whether you started at the bottom of the map and worked your way upwards or vice versa, as you could cross the bridges in either direction. But in some situations the direction you take is very important – for example in a traffic system with one-way streets. Then simple lines are not enough; they need arrows to show you which way you're allowed to go. In Manhattan, for example, nearly all the streets are one-way. If you want to think about driving through New York in mathematical terms, you have to take that into account. As a graph, the street plan of Manhattan would look something like this:

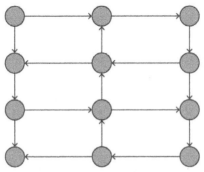

Street plan of Manhattan

Here the circles are intersections between the one-way streets, indicated by the arrows. You can see that you can get stuck at the intersection at the bottom left, as there are no lines leading away from it. This graph doesn't work as a street plan, at least not for anyone who always obeys the traffic rules.

If you delete the intersections in the left column, you can get to and leave all the intersections. With an even number of intersections, horizontally and vertically, the graph works out. Then, as you can see in the graph, you get a full circuit. The street plan doesn't work because the circuit on the left isn't complete. That's why you can't get to the intersection on the top left or away from the one at bottom left. The intersections at top and bottom right are no problem, because the circuit is complete. A mathematician can easily show a town planner that this always works, so that the planner can save time thinking about how to design a practical street plan.

Google Maps also needs arrows in its graphs, of course. To calculate a route the system must know whether a street is one-way or not. And, just as importantly, it must also know whether congestion is affecting traffic in both directions. If there's a tailback on one side of the motorway and not on the other, Google Maps only has to increase the travelling time for the people on the congested side. If you're lucky enough to be going the other way, your travelling time will stay the same. Arrows can be very useful in these situations: all the computer has to do is put an arrow in the direction of the tailback and it will be included in the calculation.

In Google Maps, then, the numbers and arrows represent travelling times and roads. As you will probably remember from Chapter 1, those two factors are all you need to calculate a route without having to look at a physical map. We saw a simple calculation to find the shortest route: the computer follows all possible routes, in order of length, until it finds the shortest one to the chosen destination. Before it finds that route, it has also followed all those that are shorter but go to the wrong destination. This calculation is known as Dijkstra's algorithm.

In the illustration, Dijkstra's algorithm is used to find the best route from the star at bottom left to the cross at top right. To make it clearer, the graph is shown in blocks rather than circles. You can imagine each block having four arrows in the direction of the adjoining blocks. In other words, you can only travel horizontally and vertically from one square to another. The dark row of blocks represents a river or some other obstacle where you can't drive with your car. Each numbered block is a place where the algorithm has looked to see if that is where you want to go, and the number stands for the number of steps in the route. The light-coloured blocks show the shortest route that the computer has identified among all the possible routes.

Graph showing all possible routes with fewer than twenty-three steps calculated by Dijkstra's algorithm*

* Image © Amit J Patel, redblobgames.com

Dijkstra's algorithm has calculated the route very systematically, as always. The computer looks first at all the routes that are one block long. They are marked with a 1 in the figure. It then moves on to all routes marked with a 2. It takes a very long time to get to the cross (twenty-two blocks away), because there are a lot of routes shorter than twenty-three blocks. The computer calculates all of them, even though they don't end up at the destination. And that's the problem with this algorithm: the computer has to make a lot of calculations before it finds the right route.

The more roads there are and the further away the destination, the longer it takes to calculate the route. That's why Google Maps doesn't use this algorithm. As is often the case with such companies, the exact programme Google uses is a secret, but we can make a reasonable guess, because we do know what methods are popular in finding routes. One of these is the A* (A star) algorithm. A* is similar to Dijkstra's algorithm, in that it also calculates all the shortest possible routes. But A* includes an estimate of the distance to the destination.

It's not difficult to make such an estimation. Although the computer can't see the whole graph, with a little extra information it can get quite a long way. Google Maps, for example, knows the coordinates of your starting point and destination. The average distance between degrees of latitude is 111 kilometres, while that between degrees of longitude varies from about 111 kilometres at the equator to near-zero at the poles. So, once the computer knows how many degrees of longitude and latitude there are between the start and the end of the route, it can estimate the distance between them and how long it will take you to reach our destination. That estimate takes no account at all of the number of roads, speed limits, traffic congestion, etc. Google therefore undoubtedly uses a better method to make its estimates. We don't know exactly what that is, but the idea behind it won't be very different from the A* algorithm. Google makes a smart estimate of how long a route will take before starting to calculate it mathematically.

The mathematical trick behind the A* algorithm is that it doesn't only look at the distance that has been covered, but also estimates the distance still to go. It then only considers routes where the sum of those two values is as low as possible. That can make a lot of difference. The illustration shows how the A* algorithm works out the route calculated above using Dijkstra's algorithm. As you can see, the computer looks at far fewer possible routes.

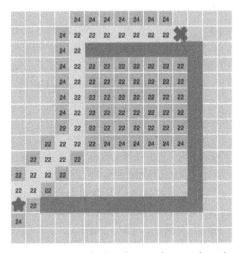

The same route calculated using the A* algorithm*

* Image © Amit J Patel, redblobgames.com

In this example, the A* algorithm's estimate of twenty-two blocks was very accurate, coming up with the actual shortest route. It reached that estimate with a piece of maths similar to that used for calculating with latitude and longitude: it deducted the coordinates of the starting point (three from the bottom, one from the left) from those of the destination (fourteen from the bottom, twelve from the left). The resulting sum was $(14 − 3) + (12 − 1) = 22$. Intermediate estimates of the remaining distance to the destination can also be made using this method.

The computer first looks at a lot of routes that don't work. For example, along the river, where it might be able to get to the destination more quickly if there was a bridge. And yet, A* is still clearly quicker than Dijkstra's algorithm. There are no blocks in the bottom right of the figure, in completely the wrong direction. That's because that area is difficult to reach and the algorithm has to take a lot of steps from the starting point to get there, so the computer has estimated that it's too far from the destination. The sum of those two distances is very high for routes that are in completely the wrong direction, so A* doesn't take them into account.

A* will always find the shortest route to the destination, as long as its estimate of the distance is not higher than the actual shortest route. That will never happen with the method used in this example, but, if A* uses a more complicated method than simply deducting the coordinates, there is a risk that it will come up with a higher estimate and will not find the shortest route. The algorithm might then follow a convoluted route and unexpectedly reach the destination, which was actually a lot closer than the high estimate had suggested. For example, while exploring the options it may have taken a snaking route through a lot of side streets (unaware that they were side streets) and happened to hit on the destination. The computer couldn't have predicted that because, according to the high estimate, it was nowhere near it.

Despite this one downside, A* is considerably better than Dijkstra's algorithm because the computer has to make far fewer calculations over long distances. There are also other mathematical tricks to speed up the process. For example, the computer can calculate the route in both directions at the same time, from the starting point and the destination, until the two routes come together. It calculates the first step from the starting point, then the first one from the destination, followed by the second step from the start, and so on. Using A* it estimates the remaining distance in both directions. With smart programming, a computer can even calculate efficient routes through the entire North American road network.

An experiment with that network showed just how enormous the difference is between the two methods. The whole network comprises 21,133,774 circles (intersections) and 53,523,592 connecting lines (roads). Dijkstra's algorithm passed through an average of 6,938,720 of those circles to find a route, while a two-directional A* algorithm making estimates to reduce the size of the graph found the shortest route by looking at 'only' 162,744 circles.

Making those preliminary estimates is very important, and the most creative innovations are currently being made in that area. The widely used 'highway hierarchies', for example, simplify the graph so that a normal computer can calculate the shortest route through the European road network, with 18 million circles in the original graph, within a thousandth of a second. The name says it all: long routes are probably the shortest on motorways.

It would take a lot longer to drive from Paris to Rome if you only used country roads, so a smart computer ignores them. It only looks at the smaller roads that connect your starting point and destination to the motorway network.

The computer doesn't know in advance which of the arrows are motorways. The main challenge for the 'highway hierarchies' method is to identify motorways without someone having to do it beforehand. It can do that completely automatically by looking at which roads the computer comes across most frequently in the shortest routes in the original graph. Country roads will not appear very often in the shortest routes over long distances, so the computer ignores them, leaving only the more important roads. The idea behind the 'highway hierarchies' method, that motorways are roads that are used by a lot of people, means that the computer can ignore millions of unimportant circles and arrows and therefore have far fewer routes to calculate.

Going back to the route from Paris to Rome, the computer first looks at how to get from the starting point in Paris to the closest motorway and then, since it is calculating the route from both ends, does the same from the destination in Rome. Once it has found the motorways, it can ignore all other roads and continue to calculate until the routes from both ends meet on the motorway. By ignoring all roads except motorways, the calculation is made a lot simpler. In this way, a series of mathematical ideas – from estimating the length of your journey on the basis of your coordinates to recognising motorways by the number of journeys made on them – have enabled us automatically to calculate routes over enormous distances.

Journeys through the internet

We come across graphs like this every day, not only when we use Google Maps but also when we're not on the move. Google uses them every time you look something up. The hits you see are largely based on Google making a journey of its own – through the internet. These journeys have made search machines a lot better. As I said in the Introduction, before Google arrived search machines couldn't even find themselves!

Google's founders solved the problem of how to find the most important internet pages with mathematics. They were the first to see the internet as one big graph, with web pages connected to each other by hyperlinks. Within Wikipedia, for example, you can travel though the internet by jumping from page to page. And if you keep coming back to the same page during your journey, it must be an important one because it ranks high on the list of hits. Google therefore spends a lot of time travelling through the internet to see where you end up when you make a search. And that will much more often be Wikipedia than an obscure website showing only a dated picture of Bill Clinton.

Google does that, of course, using a mathematical calculation. And that's a good thing, because then it can be much more certain that it is tagging the right websites as important. If you just travel though the internet at random, there's a good chance that you'll get stuck somewhere, perhaps in a group of websites on conspiracy theories. They all have links to each other, but are not a more important source of information than Wikipedia. Google's calculation also makes sure that you can see the difference, which is not always possible if you surf the net at random. That way most searches won't end up with conspiracy theories, but with accurate information. A recent study found, for example, that Google ranks news sites that are more factually accurate higher.

Imagine that the graph below shows the whole internet. The letters in the circles represent websites. B might be Wikipedia, for example, a reliable source of information that a lot of other sites refer to. The numbers indicate how important each site is according to Google; these are the scores that Google has to calculate. A high score means that a website is important, while a low score means that you really need to know about the site if you want to visit it. At least, according to Google.

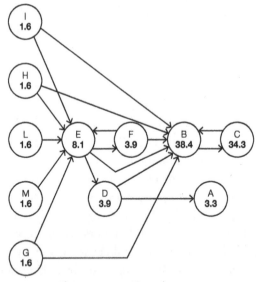

The internet as Google sees it

You calculate these numbers by pretending that you really do take a journey through the whole internet. You go from one website to another via links, the arrows on the graph.

You read something on website I and then go to site E. From there, via F, you end up at B. In this example, practically all routes lead to B (Wikipedia) and it therefore gets a high score. You soon find yourself on one of its pages, and with good reason.

Wikipedia itself refers to another website, C, as a source of further information. Although C is linked only to Wikipedia, it gets a high score because – despite all the complaints – Wikipedia is now quite reliable. D, which only one other site refers to, has a much lower score. So it's not only important how many sites refer to the page you are looking at, but also how highly those sites score.

Is it getting any clearer? Let's say that you have to choose between two websites: a Wikipedia page about 9/11 and one promoting a conspiracy theory surrounding the attacks. According to Google, you should first look at the number of links. Wikipedia has a lot, but the people behind the conspiracy theory have spent a lot of money making sure that their site has more. So a lot of websites talking all kinds of nonsense that hardly anyone ever visits have a link to the conspiracy site. Does that mean that Wikipedia is now suddenly less important? Not at all. Google doesn't want people spending a lot of money to make sure that their conspiracy theory site is the first hit in a search for 9/11: Google users want to see the most reliable information, not information on which the most money has been spent. Google can partly ensure that happens by taking into account how important the sites are that refer to a specific page. You can't pay the BBC to place links to conspiracy theories on its site, which makes links on the BBC site worth a lot more to Google than links which a different site actively sells.

This is not, however, how we really use the internet. When did you last visit fifty websites one after the other by clicking from link to link? Mostly, if you want to go to Facebook, for example, you just type in the URL. It's too much work only to access well-known sites through links. Google doesn't always do that either. Sometimes, while making its calculation, it will go straight to a site. Likewise, you might suddenly jump from Wikipedia to Facebook (in the illustration that could be from B to E), because you can't resist seeing whether your friends have posted something new. You've entered Facebook in the address bar, rather than following links on websites – arrows in the graph – finally to get to Facebook. In the case of the calculation, the probability that the computer uses a URL rather than a link (jumps to a random circle rather than following an arrow) is about one in six. This doesn't perfectly mimic our behaviour, but a one-in-six probability of the computer typing in a URL gives an accurate picture of the relative importance of websites without slowing down the calculation too much (the lower the probability, the longer the computation takes).

In the end, it is actually nothing more than a large puzzle which has to be filled in with numbers. If you go to website C via B, C will get a higher score because the incoming link makes the website more important. If you go back to B from C, then B also gets a higher score because one of the links to B has become more important. And that raises C's score again, and so on. Luckily, the scores don't keep rising endlessly. It can be proved mathematically that, at a certain point, they don't change any more. Google usually stops after about fifty calculations. In other words, for every website that can be found using its search engine, it computes a score fifty times! Only then do the scores no longer change.

So the maths works like this: you travel through a graph of links, entering a URL now and again to visit a different website at random. The most important sites are those with links to a lot of other important sites – in other words, the pages that you often come across while taking your journey through the internet. And that proves to work very well, not only for websites but also in other situations, for example finding out which films you might like.

Watching films with a graph

Netflix uses the same calculation to recommend new films and series. We don't know for sure, because Netflix is as secretive as Google about its algorithms, but its computers probably also use a graph through which you travel to arrive at its recommendations. You might look for something completely different, perhaps because you saw a poster of a film or a friend was very enthusiastic about it. Netflix uses your choices to try to understand what your taste is. It categorises you in a few groups and makes recommendations based on that categorisation. Behind all this is the same mathematical calculation that Google uses. Think back to Chapter 1, where we looked at what happens if you watch *Iron Man*. Netflix uses what it knows about all its other subscribers, what they watched after seeing *Iron Man* and how many people watched *Iron Man 2* after watching *Iron Man*. If there were a lot of them, it's a good recommendation and it gives *Iron Man 2* a high percentage, showing that it is similar to the films and series you've already seen. *Blue Planet* doesn't get such a high score because there are far fewer people who like to watch action movies *and* nature documentaries.

Mathematically, there's hardly any difference between Netflix and Google. Netflix's algorithm imitates people's behaviour. It assumes that you will largely watch films that a lot of other people with your tastes have also watched;

just as Google assumes that websites are important if a lot of other important websites refer to them. You will like a film if it is similar to a lot of other films you like. And if you feel a little more adventurous now and then and watch something completely different just to see if you like it, the mathematical calculation doesn't simply go in a straight line to a similar film or series, but jumps to a completely different place in the graph.

So we see that a piece of maths developed to find the most important information on the internet can also be used to find films and series that suit users' tastes. Both take a lot of calculating capacity. Netflix has a huge number of films and series, and they all have to be given a score. And, just as with Google, that score has to be recalculated over and over again, and for each individual user. Its calculations try to make sure that the films and series it offers you suit your preferences.

It doesn't always work out that way, however: Netflix can't recommend a film or series for you that is completely different from what you've ever seen before. In other words, it can't really surprise you. It calculates its recommendations by looking at films that are most similar to films you've seen before and not at something completely different that you might also like. The maths can't do that: it has no understanding at all of films.

Using maths to treat cancer more effectively

It's not only large internet companies that like working with graphs. Hospitals use them too, for example to predict how effective a specific treatment for cancer will be. That varies from patient to patient, partly because of differences in genes; but it has proved possible to predict those differences quite accurately with the same calculation used by Google and Netflix. Before they used the calculation, doctors were correct in 60 per cent of cases. When they started using graph theory in a 2012 study, they had a success rate of 72 per cent right from the start. That's an enormous improvement for all those patients who would otherwise have lost valuable time on treatments that may not have worked.

Cancer-treatment predictions are based on a small group of genes. Before maths was used, selecting that group was a matter of guesswork. Each researcher chose their own group of genes to focus on, which was often totally different from the genes their colleagues looked at. No one knew exactly which genes were the most important and, as there are so many, it was almost impossible to make a reasoned choice. What makes it even more difficult is that researchers are looking for genes that behave differently after a specific treatment, but some do that by influencing the behaviour of other genes without changing visibly themselves. Consequently, important genes can conceal themselves from researchers. It's a tough job identifying which genes change significantly after a specific treatment.

As searching through an enormous quantity of information to find something specific is exactly what Google and Netflix do, a group of researchers came up with the idea of applying the same calculations to genes. So they devised a graph based on experiments on how genes change their behaviour and how they influence each other. The circles were the genes and the lines between them indicated how much they affect each other.

There was one small difference from Google and Netflix: the researchers didn't start by giving all the genes the same score. Their starting scores were based on other research that associated genes with a patient's chances of survival.

A very active gene, for example, can help combat cancer and is therefore given a high starting score, meaning that it is important for the doctors to look at it. The graph then works exactly the same way as with Google and Netflix: a computer goes from gene to gene to see how the scores change when the effects of the genes on each other are taken into account.

By continually recalculating the scores, a few genes eventually emerge that are most important for the patient's chances of survival and how they will respond to the treatment. The algorithm therefore processes all the information on the importance of genes and their influence on each other and finds exactly those genes that, directly or indirectly, are the most significant in treating cancer. The maths creates a complete overview of all the knowledge on genes much more easily and accurately than the other methods used before 2012. In this way, even though it wasn't designed for that purpose, a piece of maths can save lives.

Facebook, friendships and artificial intelligence

As a final example, Facebook also uses graphs like this to make calculations: not to categorise information, but the friends it recommends. After all, Facebook knows who is a friend with whom, and constitutes an enormous graph showing each user and linking them to their friends. By starting with you and travelling through the graph, Facebook can work out who you might also come across in real life. There's a good chance that you might meet someone with a lot of the same friends as you at a party. The same applies to someone who has the same friends as your friends. That's a little more difficult to quantify, but let's say there are about twenty. There's a chance that one of your friends might link you to one of these intermediate friends. In other words, Facebook not only knows whom you know, but whom you might get to know.

If that worries you, it's not as alarming as all the other things that Facebook knows about you: whom you call, what websites you visit, etc. There are many stories that show how much data Facebook tries to collect about its users – and about people who have never themselves opened an account – all so that it can analyse this mass of data with its graphs. Facebook doesn't order its search results the way Google does, but uses what are known as 'neural networks'. These networks are what make practically every application of artificial intelligence possible, from speech recognition to filtering spam to medical diagnoses. They are also used to optimise targeted advertising. This is how Facebook categorises its users into groups like 'people who will probably buy a Mazda within 180 days'.

Neural networks, together with colossal quantities of data, enable Facebook to know what you are likely to buy before you've even decided yourself. The idea is that graphs don't only allow us to calculate which circles are most important: we can also use them to imitate our brains. Instead of neurons that are linked together and send signals to each other, the networks are graphs with circles that exchange numbers through the arrows connecting them. Information is entered at one end and comes out as a prediction at the other: for example, information about you as a Facebook user predicting what ad categories are best tailored to what you like. These graphs are not unchanging puzzles where you try to fill in the right numbers, but a dynamic whole which can be used to make predictions about something completely different.

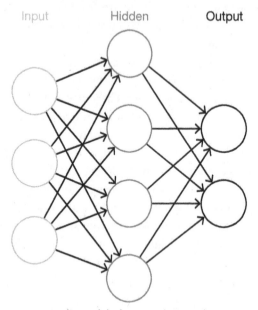

Scale model of a neural network

As the illustration shows, a neural network is a graph, but the circles play a different role. The left column is the input: just as with our brains, that is where the information goes in. With our brains that might be a photo or a face, but in these networks it's in the form of numbers between 0 and 1. Those numbers – the information – are then processed in the middle section, the four circles. A 1 in the input circle on the top left can become a 0.5 in the top circle in the middle, as every input number that leaves that circle is divided into two. The arrows change the numbers, just as the connections between neurons change information. The connections are not all equally strong, so that some neurons influence each other much more than others – they might multiply an input number or divide it by even more than half. Through a graph, the algorithm imitates the workings of the brain.

After a lot of intermediate rounds, during which the input is changed through these connections, the information arrives at the circles in the right column, the output. If the input was a picture of a face, for example, the two circles on the right could stand for the question of whether it is a man or a woman. If the computer is certain that it is a man, the circle for 'man' gets a 1 and the circle for woman a 0. How the computer arrives at this result – in other words, how exactly the input is transformed into the output – is often unknown.

In many cases, the computer itself devises the intermediate steps. It modifies the graph as it solves problems. This is the 'training phase', during which the computer gets better at solving problems by practising on photos to which it already knows the answers. That is why we often say that computers can 'learn'. They change the value of the link between the circles, i.e. the numbers next to the arrows. If the input circle at top left stands, for example, for the length of the hair of the person in the photo, that might at first not be so important to the calculation. The circle and the arrows leading from it will then have a very low score. While it is 'practising' the computer may discover that hair length is an important factor, and assign a higher number to the arrows leading from the 'hair circle'.

That requires masses of data. Facial recognition needs a lot of photos, and we have to know for the training if each photo is of a man or a woman. The computer starts off quite randomly. It doesn't recognise anything in the first photo, but it does produce an answer which is then compared with the right one, which in the training phase is used as a benchmark. On the basis of that comparison, the computer makes a number of changes before moving on to the next photo. If it repeats this procedure often enough, a computer will almost always come up with the right answer eventually.

A computer, for example, finally beat a human at Go. For a long time the board game was too difficult for computers, but the successful computer used an enormous graph to play millions of games against itself before taking on its human opponent. It used the information on how each game was won to modify the graph. In doing so, it taught itself the rules of the game and how to play as well as possible. Today, a computer only needs three days to learn the game well enough to beat the reigning world champion.

Facebook uses the same idea to find out whether you are likely to buy a Mazda. Intelligence services want to use it to identify criminals and terrorists. China is setting up a social-credit system with it, giving every member of the population a score depending on their behaviour. And there are many other possible applications, some of them quite alarming. Computers can, for example, determine a person's sexual preference from a photo. The predictions aren't perfect yet, but they are possible. And that means they're open to abuse.

The Cambridge Analytica scandal is a good example. This British-American consulting company used data from Facebook to predict political preferences, including what messages would appeal best to which people. More specifically, they worked out how to persuade people to vote for Donald Trump. The company also worked on the campaign for 2015 US presidential candidate Ted Cruz, which was not a great success. We will never know what impact the work of Cambridge Analytica had on how people voted, but what is overly clear is that the company should never have had access to all that data in the first place – and that it could do a lot with it, thanks to graph theory.

Graphs in the background

As we have seen, graphs crop up everywhere. Not directly, as with statistics, but in the background. As with integrals and differentials, we don't need to know anything about graphs to use a navigation system, or Google or Netflix. So is it important to understand something about graph theory? I think so, because the way graphs are used can have an enormous impact on our lives.

At the beginning of this chapter we saw how Google Maps uses graphs to calculate the quickest route to your destination. Like integrals and differentials, such applications don't change our lives that much. They make some things easier – you no longer have to be able to read a map, for example. But they don't get us asking more serious questions, such as whether we actually want them or not. Of course, we want a way of getting from A to B as quickly as possible. If graph theory makes that easier, that's fine. As a user, you don't really need to know how it all works.

But it's a completely different matter if Google, Facebook and other companies use graph theory to order information or make decisions with neural networks. Then it's very useful to know something about it. Take, for example, intelligence services that suddenly want access to a lot of personal data. What are they going to do with it? What information can they find in it? What steps are controlled by humans, and what happens if humans are not involved? Only someone who understands graph theory can really answer questions like these.

There are a lot of issues on which knowledge of graph theory helps form our opinions. Google and Facebook often manoeuvre their users into a 'bubble' where they are mainly fed information that confirms what they already think. As a user, you have to make a special effort to find alternative ways of thinking. Can't Google and Facebook do something about that? They have access to all those other standpoints. They're online as well, so why don't we get to see them? Why can't I actively make clear that I want to see things from another perspective too? The answer is simply because the maths used by Google and Facebook isn't designed to do that. They can't 'just' adapt their algorithms so that we get to see completely different things that still have something to do with the subject we are interested in.

As we have seen, for Google and Facebook the most important information is information that is easiest to find – in other words, that most closely resembles the information you are looking for. Just as Netflix can't recommend a film which is completely unexpected but perfectly suits your viewing preferences, Google can't offer information which falls outside the specific terms you have entered in its search bar. Filtering fake news is not as simple as it sounds, since mathematical calculations can't 'see' what is on a website. Of course, mathematicians are working hard to make that possible, but it won't happen overnight. The maths they use now doesn't make that easy.

Fake news and concerns about privacy and artificial intelligence have become important social issues, and they are all based on the possibilities and limitations of graph theory. That's why it's so important to have at least some understanding of this area of mathematics. If you want to voice an opinion on the great social issues facing us today, you need to know what underlies them, which solutions are feasible and which are not. And you can't do that without understanding graph theory.

CHAPTER 8

What maths is good for

Mathematics – let me say it again – is very useful. And relevant. In our daily lives, too, though we're not often aware of it. But how it is possible that maths works so well? I asked that question in Chapter 2, where we also saw that it doesn't matter how you think about maths – as an abstract form like Plato with his cave, or as a large fabrication like the stories of Sherlock Holmes. In both cases, the usefulness of maths is not immediately clear. If it's so abstract, how can maths have anything to do with the reality around us?

With questions like these, it often helps to keep the maths as simple as possible. So how are numbers useful? We started using them in the distant past to keep a more accurate track of quantities. We could do that because the nature of numbers allows them to be used in all kinds of situations. Positive whole numbers are special and simple: you start with 1 and just keep adding another 1. So 2 is nothing more than the number that comes between 1 and 3.

When we count, we use 1 as a kind of box that we put the first thing in, 2 as a box for the next thing, and so on. The sequence that you put things in keeps them apart, whether they are loaves of bread, sheep, coins or whatever. But it doesn't work for everything. Try counting piles of sand. Make one on the ground, then make another one next to it.

There's a good chance the two piles will collapse a little and run into each other. Then, rather than two separate piles, you'll have one again, only slightly bigger. One pile of sand plus another pile of sand still equals one pile of sand. Does that mean $1 + 1 = 1$? Not at all, it's just that numbers don't work with piles of sand because they're not separate units. You can solve that by using a unit of quantity. With sand, for example, you can use litres. Then one litre plus one litre is two litres, even if the sand all flows together into one big pile. Units put things into a form that enables you to use numbers to keep track of quantities. In other words, numbers have a very rigid structure, which is very useful because we come across that same structure in all kinds of places around us. That doesn't mean to say that that you can do anything you like with them, because it's still difficult to count piles of sand.

Back to our question: how useful are numbers? They help us structure what we encounter all around us. They are useful and easily applicable because they focus our attention on that structure, ignoring all the details which are not important at that moment. That's why maths is different from stories about Sherlock Holmes. The stories offer a reasonably good description of reality: so you can learn something about London, for example, in the time of Sherlock Holmes. But what they don't offer is the abstract. There are no structures in the stories that focus your attention on a specific property, as numbers do for quantities.

Mistakes in maths

So maths is a perfect way of ordering the things we see around us. That's why you can use it to understand quantities, for example. It sounds good, but it's not always that simple. Once we start using more complex maths, it quickly stops reflecting reality that closely and mistakes soon creep in. Google's algorithm, for example, assumes that every link from one website to another is positive, that none of them take you to a site that is full of nonsense or misinformation. You don't want pages like that to appear in your search results, but the maths can't tell the difference: it simply gives the site extra points for the link. Just as Facebook can't see in its graph which people you really know and whom you've added to your friends for a joke. As far as the maths is concerned, you're equally good friends with all your contacts on Facebook.

Because the maths simplifies situations, it by no means always provides a perfect picture. Take one of the standard questions in physics: someone fires a cannon at a castle. Where will the cannonball land? The mathematical calculation gives two answers, a positive and a negative one. The cannonball will either land against the castle wall 100 metres in the direction you have fired (the positive answer), or 100 metres in the exact opposite direction (the negative one). The latter is obviously nonsense – a cannonball never goes in completely the opposite direction of the cannon.

With numbers, it's easy to say that maths is useful because it perfectly orders the situations around us. That's true, as long as you're careful about what you are counting. If you make it a little more complicated and differences creep in between the world and the mathematical structures, the mathematical calculations will come up with something different from what happens in reality. Yet even in those cases, there are sufficient similarities to say that the maths helps. Since we know full well that a cannonball can't fly in the opposite direction, we can still use the calculation; we just discard the negative number.

So how exactly does that work? What similarities do there have to be for the maths to be useful? How many mistakes can it contain? We don't know. Philosophers are busily discussing these questions. Waiting until they've agreed is never a good idea, so for the time being it's enough to say that the similarities help to explain how the maths works in practice. Maths structures the world around us in ways we may not have noticed otherwise. It makes it easier to forget details and concentrate on the practical problem we are facing.

Is it all coincidence?

Similarities to reality – that's what's important in determining how useful maths can be. But where do the similarities come from? Did they just appear from nowhere, or have mathematicians worked hard to make maths something we can also use? There's no clear answer to that question yet either. Look at what mathematicians themselves found important. Archimedes, who made all kinds of practical discoveries, thought his theorem on spheres, cylinders and cones the most important of them all. But that was the very discovery that clearly had no practical use. What does it matter how much of a cylinder you have to remove to get a cone? You can find that out for yourself by doing it.

Mathematicians often don't concern themselves with how their discoveries can be applied in practice. That's why it seems almost coincidence that maths is so useful; perhaps not for numbers and for geometry, but certainly for the areas of mathematics we have looked at. As we saw in Chapter 3, arithmetic and geometry began as a way of solving practical problems. When people started living in increasingly large communities, it presented serious administrative problems.

City states had to find more efficient ways of collecting taxes, keeping track of food supplies and planning for the future, and numbers helped them solve those problems.

But those numbers evolved only slowly. In Mesopotamia they had stone tokens, a convenient way of keeping track of how many goods you had by taking the same number of tokens with you. In the course of time the tokens gave way to marks on clay tablets, which were easier to carry around than a pile of stones. In short, we started using numbers because they were useful. It was not coincidence; the first mathematical sums were extremely practical. The maths was useful because it solved a difficult problem.

A few centuries later, the picture is a little less clear-cut. Mathematicians in various cultures started studying 'useless' problems. They wanted to solve them more out of curiosity or to increase their status than for their usefulness. And that's still the case: we have most respect for that 'useless' maths. We largely remember the Greeks for their highly abstract mathematical work. Almost no one has heard of Eupalinos, the man who dug the tunnel, but everyone knows the name of Pythagoras, the ultimate mathematician.

No matter why mathematicians developed abstract ideas, they can be applied in practice. Pythagoras' theorem is a useful way of finding out whether you have a right-angled triangle, and much of Archimedes' work has many direct practical applications. As does more difficult maths, like calculus, probability theory and graph theory; strangely enough, if you look at history they were often not developed by coincidence either.

Newton and Leibniz, for example, knew straight away that calculus would become important. Newton immediately used it in his own physics. Although it was difficult at first, they were still able to apply their mathematical theories directly, because the idea behind them was actually very simple: they wanted to study change. And of course we see change all around us. Newton saw it in his maths too, by imagining he was walking through a graph. That made the idea a little more abstract, but no less relevant.

Of course, a method to calculate change is applicable in practice. And that means its development wasn't as coincidental as it may appear. Similarly, probability theory started with games that had to stop before they ended. That seems to have nothing to do with opinion polls, diseases or crime figures, but indirectly it has. Mathematicians were asking themselves how to calculate things that you don't know for certain, how to deal with uncertainty in an exact way.

We encounter uncertainty in all kinds of situations. So if you know a way of calculating that uncertainty mathematically, you can use it to study the world around you. Not that it was simple to apply probability theory. It literally took centuries before we could hold polls and calculate their accuracy mathematically. The point is that these applications are not coincidental; mathematicians became interested in uncertainty and worked on something that could ultimately be used to study concrete uncertainties around us. Useless games or not, the potential was already there in the subjects they chose to study.

That even applies to graph theory. Euler initiated this field of maths on the basis of a puzzle: the seven bridges of Königsberg. The puzzle itself had little practical value: what use was it to know that it's mathematically impossible to walk though the city centre and cross each bridge only once? The underlying idea is also not immediately visible: taking a walk seems to have little to do with route-planners or search engines – not until we look at the problem a little more generally, which is easier to do with hindsight. Euler studied a network, a way in which different places are connected to each other. And later, we encountered similar networks in other situations.

That's especially true today. Social networks are an obvious example, but there are many more. Traffic networks are also easy to study with graph theory. And train and metro networks, so that timetables can be drawn up. And, as we have seen, networks of films and series, or of genes that affect each other's behaviour. Graph theory is the general study of networks and their properties, so again it's no coincidence that it can be applied widely in practice.

The abstract work of mathematicians is thus often inspired by things that we encounter on a day-to-day basis. That's why it's no coincidence that these areas of mathematics can be used to better understand the world around us. There's good reason why maths is useful.

Maths helps

So we've already looked at two big questions: what makes maths useful, and is that pure coincidence? But why should we want to use maths in this way? As I said earlier, maths doesn't allow us to do anything we couldn't have done without it. Look at the Pirahã and the other cultures in Chapter 2. They can work with quantities, shapes, social groups, change, and so on perfectly well without using maths. If someone were to show them how to build machines, I'm sure they could follow all the necessary steps and do it themselves. After all, the maths isn't in the machines or the buildings. People can do a lot of things without the help of mathematics; it's just a lot more difficult.

The similarities in structure between mathematical ideas and reality mean that maths makes it easier to solve practical problems. Maths simplifies reality. By focusing only on the structure, you don't have to keep all the details in your head. We can't see the difference between twenty-one and twenty-two loaves of bread until we arrange them neatly in two rows and see that one of the rows is slightly longer. And that's essentially what maths does for us.

Take the weather, for example. For a long time we predicted it without using maths, and we still can. We simply study the current weather very closely and think about how it might change. If the wind is coming from the east and we see a lot of showers in that direction, we can be pretty sure it's going to rain. But it's difficult to keep track of all those small differences and changes; so much changes so quickly that we simply can't do it. We certainly don't have the time. You could write everything down in a big book and spend 100 years figuring it all out, but that's no use to anyone.

Maths helps us to focus on the most important aspects of the weather, like air flows and how they change over time. Of course, it helps that we can leave these calculations to a computer, otherwise it would still not be feasible to use formulas to make weather forecasts. We can only do that thanks to calculus. Without it, even a computer can't predict the weather.

So maths helps by making complex problems simpler. You apply a mathematical method because there's a similarity between the structures in the maths and in reality. Thanks to that similarity, you can ignore the details you encounter in the real world. You can make time stand still and look at all the pieces of the weather at your leisure. Or you can ignore all the differences between people and concentrate only on average income or political preferences. That makes it a lot easier to solve problems.

That's how a lot of the maths discussed in this book

works. But sometimes, mathematics can be useful for a completely different reason: it can suggest new solutions. We saw some examples of that in Chapter 1.

Maths regularly comes up with surprises in physics. The scientists Paul Dirac and Augustin Fresnel made new discoveries when their calculations produced unexpected results. Just as with the cannonball, their experiments with particles and the behaviour of light came up with seemingly crazy results, but they proved to be correct. Mathematics fits better with reality than we think; it even shows us things that we ourselves haven't yet noticed.

We don't know how that happens. It's a mystery why maths works so well. That is, if it really does work that well and we haven't just been lucky. While it's reasonably clear how maths makes problems simpler, how it helps us to find new theories, how seemingly strange results can sometimes lead to new discoveries is much less so. But it doesn't make the maths any less special.

In our daily lives, too

These new discoveries are usually made in science. Most people probably don't use maths enough to encounter strange mathematical predictions themselves. But, even though we may not use it actively ourselves, maths is useful in our daily lives because it makes the world around us easier to understand. We don't need to use calculus after we've left secondary school and, as for all those formulas we spent too much time staring at, if we're lucky we'll hardly ever see them again. Even I don't need to use them, and I'm a philosopher of mathematics. So why do I keep hammering on that, despite all this, we need to know something about maths?

Maths isn't the only thing that we come into contact with indirectly, day in, day out. How about car engines or politics? Both have a great impact on our lives. Without cars, getting around would be a lot harder for us, and without trucks and vans we'd have more trouble getting the goods we buy and use. The same applies to politics. Generally speaking, most of us have no direct contact with it and yet political decisions affect us all. Car engines and politics have an indirect but deep impact on our lives. But does that mean we have to know how they work?

In the case of a car engine, that would be crazy. Car drivers don't need to know how their cars work. All they are interested in is that they do. A different engine – say, a switch from petrol to electric – doesn't make much difference to your life. Your car will keep driving and the economy will keep going. It might be more environmentally friendly, but not essentially different.

With politics, it's different. A change from a democratic to an authoritarian regime will certainly affect everyone. Smaller changes, like a new law being passed or not, also impact on our daily lives. That's why we all learn at school how the political system works. Although politics may sometimes seem remote from your daily life, it's clearly important to know something of how it works. You won't be confronted with it on a daily basis, but it's no less important to know what's going on.

The same applies to mathematics, though there is a difference between different areas of maths. Very theoretical disciplines, like set theory, have hardly anything to do with everyday life and that's why I haven't looked at them in this book. But there are also big differences between the areas of maths that are used a lot in practice. Integrals and differentials are very important, but have more in common with car engines than politics. It wouldn't be a great problem if mathematicians came up with another way to calculate change. There are even different ways to use integrals and differentials, but it doesn't matter which method is used. They all produce the same weather forecasts, the same buildings and the same predictions for election programmes. The main thing is that the method works. We don't need to think too much about how.

Integrals and differentials are used in so many places that it can do no harm to understand a little about them. We come across them in many occupations and they've played a very significant role in the development of today's society. As I said in Chapter 5, you can compare them with history. It's useful to know how the world around you came into being, why things are now as they are, as it gives you a better perspective on society. You can look at integrals and differentials in the same way. Calculus is one of the most influential ideas in history. So it's logical to learn something about it even if the details of the calculations are not directly important in our everyday lives.

Statistics, on the other hand, do have a great impact on how we live from day to day. How improvements in average income are calculated can make a great difference to the result, and therefore to our image of society. The same applies to polls, data on salary gaps between men and women, and the results of scientific research. Statistics can help us tremendously by making large quantities of data accessible and showing us correlations that we may not have noticed. The problem is that statistics can easily be used or manipulated to distort our view of the world.

What methods are used, how a poll is conducted or what an average is based on can affect how we see the world and how we make decisions. To form our own opinions, it's therefore important to be able to look at statistics with a critical eye, just as we need to be able to look at politics critically and not believe everything politicians tell us. And knowledge of maths is essential in order for us to do that. Not to make calculations ourselves, but to understand where it can go wrong. Statistics are invaluable in our daily lives.

Lastly, there is graph theory. That also has an enormous impact on our lives, because companies like Google and Facebook use graphs to determine what information we get to see. That's what makes this area of maths much more invasive than statistics: if Google changes the way it uses graphs, for example, it can mean that you see completely different information, that you might be misled or even have hardly any access to information at all. We already see that in the information bubbles that ensure that people mainly come into contact with others who think like them and have the same preferences.

Graph theory shows how we obtain information through websites like Google. At least as important is that we see what happens to information we provide in return. What can Google, Facebook and other internet companies do with all the personal data they collect? Who looks at it and what parts of the process are fully automatic? These are questions that people are very concerned about and, if they want really good answers, they need to understand the maths behind them. Then they can talk about what is possible and what isn't, how artificial intelligence really works and where the dangers lie.

Who has enough time to do all that alongside everything else? To check every figure you come across, to keep up to date on all the latest developments in artificial intelligence and still lead a normal life? No one does, and they don't need to. You can get a long way just by understanding the basics. That's enough to enable you to look critically at the surprising results of a study or an opinion poll. And to think more actively about what data intelligence services may and may not collect, because a little maths gives you a better idea of what they do with that data.

Mathematics, and certainly the more difficult parts of it, gives us a greater insight into the world around us. You might hardly come across any actual calculations in your daily life, but – I would say to my fifteen-year-old self – what you see all around you is what maths studies. Strangely shaped buildings, weather forecasts, polls and predictions based on masses of data, search engines and artificial intelligence – these are all things you will understand better if you can grasp the core ideas of mathematics. Especially now that the world is becoming more and more complex, we need something to make it a little less confusing. And that's what maths does – and in a way that is easier to understand than we often think.

BIBLIOGRAPHY

Barner, D., Thalwitz, D., Wood, J. et al. (2007). 'On the relation between the acquisition of singularplural morpho-syntax and the conceptual distinction between one and more than one.' *Developmental Science* 10 (3), pp. 365–73.

Batterman, R. (2009). 'On the explanatory role of mathematics in empirical science.' *The British Journal for the Philosophy of Science*, pp. 1–25.

Bauchau, O. and Craig, J. (2009). *Structural Analysis: With Applications to Aerospace Structures.* Dordrecht: Springer.

Bianchini, M., Gori, M. and Scarselli, F. (2005). 'Inside PageRank'. *ACM Transactions on Internet Technology* 5 (1), pp. 92–128.

Boyer, C. (1970). 'The History of the Calculus.' *The Two-Year College Mathematics Journal* 1 (1), pp. 60–86.

Brin, S., & Page, L. (1998). 'The Anatomy of a Large-Scale Hypertextual Web Search Engine.' *Computer Networks and ISDN Systems* 30, pp. 107–17.

Bueno, O. and Colyvan, M. (2011). 'An Inferential Conception of the Application of Mathematics.' *Noûs* 45 (2), pp. 345–74.

Buijsman, S. (2019). 'Learning the Natural Numbers as a Child'. *Noûs* 53 (1), 3-22.

Burton, D. (2011). *The History of Mathematics: An Introduction*, 7th edition. New York: McGraw-Hill.

Carey, S. (2009). 'Where Our Number Concepts Come From.' *Journal of Philosophy* 106 (4), pp. 220–54.

Cartwright, B. and Collett, T. (1982). 'How Honey Bees Use Landmarks to Guide Their Return to a Food Source.' *Nature*

295, pp. 560–64.

Chemla, K. (1997). 'What is at Stake in Mathematical Proofs from Third-Century China?' *Science in Context* 10 (2), pp. 227–51.

Chemla, K. (2003). 'Generality Above Abstraction: The General Expressed in Terms of the Paradigmatic in Mathematics in Ancient China.' *Science in Context* 16 (3), pp. 413–58.

Cheng, K. (1986). 'A Purely Geometric Module in the Rat's Spatial Representation.' *Cognition* 23, pp. 149–78.

Christensen, H. (2015). 'Banking on Better Forecasts: The New Maths of Weather Prediction.' *The Guardian*, 8 January 2015. Online at https://www.theguardian.com/science/alexs-adventures-in-numberland/2015/jan/08/banking-fore-casts-maths-weather-predictionstochastic-processes

Colyvan, M. (2001). 'The Miracle of Applied Mathematics.' *Synthese* 127 (3), pp. 265–78.

Cullen, C. (2002). 'Learning from Liu Hui? A Different Way to Do Mathematics.' *Notices of the AMS* 49 (7), pp. 783–90.

Dehaene, S., Bossini, S. and Giraux, P. (1993). 'The Mental Representation of Parity and Number Magnitude.' *Journal of Experimental Psychology: General* 122, pp. 371–96.

Dehaene, S., Izard, V., Pica, P. et al. (2006). 'Core Knowledge of Geometry in an Amazonian Indigene Group.' *Science* 311, pp. 381–4.

Doeller, C., Barry, C. and Burgess, S. (2010). 'Evidence for Grid Cells in a Human Memory Network.' *Nature* 463, pp. 657–61.

Dorato, M. (2005). 'The Laws of Nature and The Effectiveness of Mathematics.' In: *The Role of Mathematics in Physical Sciences*, pp. 131–44. Dordrecht: Springer.

Edwards, C. (1979). *The Historical Development of the Calculus*. Dordrecht: Springer.

Englund, R. (2000). 'Hard Work – Where Will It Get You? Labor Management in Ur III Mesopotamia.' *Journal of Near Eastern Studies* 50 (4), pp. 255–80.

Ekstrom, A., Kahana, M., Caplan, J. et al. (2003). 'Cellular Networks Underlying Human Spatial Navigation.' *Nature* 425, pp. 184–7.

Epstein, R. and Kanwisher, N. (1998). 'A Cortical Representation of the Local Visual Environment.' *Nature* 392, pp. 598–601.

Everett, D. (2005). 'Cultural Constraints on Grammar and Cognition in Pirahã: Another Look at the Design Features of Human Language.' *Current Anthropology* 46 (4), pp. 621–46.

Ezzamel, M. and Hoskin, K. (2002). 'Retheorizing Accounting, Writing and Money with Evidence from Mesopotamia and Ancient Egypt.' *Critical Perspectives on Accounting* 13, pp. 333–67.

Feigenson, L., Carey, S. and Hauser, M. (2002). 'The Representations Underlying Infants' Choice of More: Object Files versus Analog Magnitudes.' *Psychological Science* 13 (2), pp. 150–56.

Feigenson, L. and Carey, S. (2003). 'Tracking Individuals via Object Files: Evidence from Infants' Manual Search.' *Developmental Science* 6 (5), pp. 568–84.

Feigenson, L., Dehaene, S. and Spelke, E. (2004). 'Core systems of Number.' *Trends in Cognitive Sciences* 8 (7), pp. 307–14.

Ferreirós, J. (2015). *Mathematical Knowledge and the Interplay of Practices*. Princeton: Princeton University Press.

Fias, W. and Fischer, M. (2005). 'Spatial Representation of Number.' In: Campbell, J. (ed.), *Handbook of Mathematical Cognition*, pp. 43–54. New York: Psychology Press.

Fias, W., Van Dijck, J. and Gevers, W. (2011). 'How Is Number Associated with Space? The Role of Working Memory.' In: Dehaene, S. and Brannon, E. (eds), *Space, Time and Number in the Brain: Searching for the Foundations of Mathematical Thought*, pp. 133–48. Amsterdam: Elsevier Science.

Fienberg, S. (1992). 'A Brief History of Statistics in Three and One-Half Chapters: A Review Essay.' *Statistical Science* 7 (2), pp. 208–25.

Fischer, R. (1956). 'Mathematics of a Lady Tasting Tea.' In: Newman, J. (ed.), *The World of Mathematics*, bk. III, vol. VIII, Statistics and Design of Experiments, pp. 1514–21. New York: Simon & Schuster.

Franceschet, M. (2011). 'PageRank: Standing on the Shoulders of Giants.' *Communications of the ACM* 54 (6), pp. 92–101.

Frank, M., Everett, D., Fedorenko, E. et al. (2008). 'Number as a Cognitive Technology: Evidence from Pirahã Language and Cognition.' *Cognition* 108, pp. 819–24.

Freedman, D. (1999). 'From Association to Causation: Some

Remarks on the History of Statistics.' *Journal de la société française de statistique* 140 (3), pp. 5–32.

Fresnel, A. (1831). 'Über das Gesetz der Modificationen, welche die Reflexion dem polarisirten Lichte einprägt.' *Annalen der Physik* 98 (5), pp. 90–126.

Geisberger, R., Sanders, P., Schultes, D. and Delling, D. (2008). 'Contraction Hierarchies: Faster and Simpler Hierarchical Routing in Road Networks.' In: McGeoch C.C. (ed.), *Experimental Algorithms*. WEA 2008. Lecture Notes in Computer Science, vol. 5038, pp. 319–33. Heidelberg: Springer Berlin.

Gleich, D. (2015). 'PageRank Beyond the Web.' *SIAM Review* 57 (3), pp. 321–63.

Gordon, P. (2004). 'Numerical Cognition without Words: Evidence from Amazonia.' *Science* 306, pp. 496–9.

Gori, M. and Pucci, A. (2007). 'ItemRank: A Random-Walk Based Scoring Algorithm for Recommender Engines.' *IJCAI-07 Proceedings of the 20th International Joint Conference on Artificial Intelligence*, pp. 2766–71.

Hamming, R. (1980). 'The Unreasonable Effectiveness of Mathematics.' *American Mathematical Monthly* 87 (2), pp. 81–90.

Hensley, S. (2008). 'Too Much Safety Makes Kids Fat.' *Wall Street Journal*, 13 August 2008. Online at https://blogs.wsj.com/health/2008/08/13/too-much-safety-makeskids-fat/

Hermer, L. and Spelke, E. (1994). 'A Geometric Process for Spatial Re-orientation in Young Children.' *Nature* 370, pp. 57–9.

Hodgkin, L. (2005). *A History of Mathematics: From Mesopotamia to Modernity*. Oxford: Oxford University Press.

Høyrup, J. (2001). 'Early Mesopotamia: A Statal Society Shaped by and Shaping Its Mathematics.' Contribution to *Les mathématiques et l'état*, CIRM Luminy, 15–19 October 2001. Photocopy, Roskilde University. Online at http://akira.ruc.dk/~jensh/Publications/2001 per cent7BK per cent7 D04_Luminy.pdf

Høyrup, J. (2007). 'The Roles of Mesopotamian Bronze Age Mathematics: Tool for State Formation and Administration – Carrier of Teachers' Professional Intellectual Autonomy.' *Educational Studies in Mathematics* 66, pp. 257–71.

Høyrup, J. (2014a). 'A Hypothetical History of Old Babylonian

Mathematics: Places, Passages, Stages, Development.' *Ganita Bharati* 34, pp. 1–23.

Høyrup, J. (2014b). 'Written Mathematical Traditions in Ancient Mesopotamia: Knowledge, Ignorance, and Reasonable Guesses.' In: Bawanypeck, D. and Imhausen, A. (eds), *Traditions of Written Knowledge in Ancient Egypt and Mesopotamia*. Proceedings of two workshops held at Goethe University, Frankfurt/Main, December 2011 and May 2012, pp. 189–213. Münster: Ugarit-Verlag.

Huff, D. (1954). *How to Lie with Statistics*. New York: W. W. Norton & Company.

Imhausen, A. (2003a). 'Calculating the Daily Bread: Rations in Theory and Practice.' *Historia Mathematica* 30, pp. 3–16.

Imhausen, A. (2003b). 'Egyptian Mathematical Texts and Their Contexts.' *Science in Context* 16 (3), pp. 367–89.

Imhausen, A. (2006). 'Ancient Egyptian Mathematics: New Perspectives on Old Sources.' *The Mathematical Intelligencer* 28 (1), pp. 19–27.

Izard, V., Pica, P., Spelke, E. et al. (2011). *Proceedings of the National Academy of Sciences* 108 (24), pp. 9782–7.

Kennedy, C., Blumenthal, M., Clement, S. et al. (2017). 'An Evaluation of 2016 Election Polls in the U.S.' *American Association for Public Opinion Research*, report published 4 May 2017. Online at https://www.aapor.org/Education-Resources/Reports/An-Evaluation-of-2016-Election-Polls-in-the-U-S.aspx

Kleiner, I. (2001). 'History of the Infinitely Small and the Infinitely Large in Calculus.' *Educational Studies in Mathematics* 48, pp. 137–74.

Langville, A. and Meyer, C. (2004). 'Deeper Inside PageRank.' *Internet Mathematics* 1 (3), pp. 335–80.

Lax, P. and Terrell, M. (2014). *Calculus With Applications*. Dordrecht: Springer.

Lee, S., Spelke, E. and Vallortigara, G. (2012). 'Chicks, like Children, Spontaneously Reorient by Three-Dimensional Environmental Geometry, Not by Image Matching.' *Biology Letters* 8 (4), pp. 492–4.

Li, P., Ogura, T., Barner, D. et al. (2009). 'Does the Conceptual Distinction Between Singular and Plural Sets Depend on

Language?' *Developmental Psychology* 45 (6), pp. 1644–53.

Lützen, J. (2011). 'The Physical Origin of Physically Useful Mathematics.' *Interdisciplinary Science Reviews* 36 (3), pp. 229–43.

Madden, D. and Keri, A. (2009). 'The Mathematics behind Polling.' Online at http://math.arizona.edu/~jwatkins/505d/Lesson_12.pdf.

Malet, A. (2006). 'Renaissance Notions of Number and Magnitude.' *Historia Mathematica* 33, pp. 63–81.

Melville, D. (2002). 'Ration Computations at Fara: Multiplication or Repeated Addition?' In: Steele, J. and Imhausen, A. (eds), *Under One Sky: Astronomy and Mathematics in the Ancient Near East*, pp. 237–52. Münster: Ugarit-Verlag.

Melville, D. (2004). 'Poles and Walls in Mesopotamia and Egypt.' *Historia Mathematica* 31, pp. 148–62.

Mercer, A., Deane, C. and McGeeny, K. (2016). 'Why 2016 Election Polls Missed Their Mark.' *Pew Research Center*, 9 November 2016. Online at http://www.pewresearch.org/fact-tank/2016/11/09/why-2016-election-polls-missed-their-mark/.

Morrisson, J., Breitling, R., Higham, D. et al. (2005). 'GeneRank: Using Search Engine Technology for the Analysis of Microarray Experiments.' *BMC Bioinformatics* 6, p. 233.

Negen, J. and Sarnecka, B. (2012). 'Number-Concept Acquisition and General Vocabulary Development.' *Child Development* 83 (6), pp. 2019–27.

Nuerk, H., Moeller, K. and Willmes, K. (2015). 'Multi-digit Number Processing: Overview, Conceptual Clarifications, and Language Influences.' In: Kadosh, C. and Dowker, A. (eds), *The Oxford Handbook of Numerical Cognition*, pp. 106–39. Oxford: Oxford University Press.

Núñez, R. (2017). 'Is There Really an Evolved Capacity for Number?' *Trends in Cognitive Sciences* 21, pp. 409–24.

Owens, K. (2001a). 'Indigenous Mathematics: A Rich Diversity.' In: *Proceedings of the Eighteenth Biennial Conference of The Australian Association of Mathematics Teachers*, Australian Association of Mathematics Teachers Inc., Adelaide, pp. 157–67.

Owens, K. (2001b). 'The Work of Glendon Lean on the Counting

Systems of Papua New Guinea and Oceania.' *Mathematics Education Research Journal* 13 (1), pp. 47–71.

Owens, K. (2012). 'Papua New Guinea Indigenous Knowledges about Mathematical Concept.' *Journal of Mathematics and Culture* 6 (1), pp. 20–50.

Owens, K. (2015). *Visuospatial Reasoning: An Ecocultural Perspective for Space, Geometry and Measurement Education.* Cham: Springer International Publishing.

Pica, P., Lemer, C., Izard, V. et al. (2004). 'Exact and Approximate Arithmetic in an Amazonian Indigene Group.' *Science* 306 (5695), pp. 499–503.

Pincock, C. (2004). 'A New Perspective on the Problem of Applying Mathematics.' *Philosophia Mathematica* 12 (2), pp. 135–61.

Pucci, A., Gori, M. and Maggini, M. (2006). 'A Random-Walk Based Scoring Algorithm Applied to Recommender Engines.' In: Nasraoui, O., Spiliopoulou, M., Srivastava, J. et al. (eds), *Advances in Web Mining and Web Usage Analysis.* WebKDD 2006. Lecture Notes in Computer Science, vol. 4811, pp. 127–46. Heidelberg: Springer Berlin.

Radford, L. (2008). 'Culture and Cognition: Towards an Anthropology of Mathematical Thinking'. In: English, L. (ed.), *Handbook of International Research in Mathematics Education*, 2nd edition, pp. 439–64. New York: Routledge.

Rice, M. and Tsotras, V. (2012). 'Bidirectional A* Search with Additive Approximation Bounds.' In: *Proceedings of the Fifth Annual Symposium on Combinatorial Search.* SOCS 2012.

Ritter, J. (2000). 'Egyptian Mathematics.' In: Selin, H. (ed.), *Mathematics Across Cultures: The History of Non-Western Mathematics*, pp. 115–36. Dordrecht: Kluwer Academic Publishers.

Robson, E. (2000). 'The Uses of Mathematics in Ancient Iraq, 6000–600 BC.' In: Selin, H. (ed.), *Mathematics Across Cultures: The History of Non-Western Mathematics*, pp. 93–113. Dordrecht: Kluwer Academic Publishers.

Robson, E. (2002). 'More than Metrology: Mathematics Education in an Old Babylonian Scribal School.' In: Imhausen, A. and Steele, J. (eds), *Under One Sky: Mathematics and Astronomy in the Ancient Near East*, pp. 325–65. Münster: Ugarit-verlag.

Sarnecka, B., Kamenskaya, V., Yamana, Y. et al. (2007). 'From Grammatical Number to Exact Numbers: Early Meanings of One, Two, and Three in English, Russian, and Japanese.' *Cognitive Psychology 55*, pp. 136–68.

Sarnecka, B. and Lee, M. (2009). 'Levels of Number Knowledge During Early Childhood.' *Journal of Experimental Child Psychology* 103, pp. 325–37.

Schlote, A., Crisostomi, E., Kirkland, S. et al. (2012). 'Traffic Modelling Framework for Electric Vehicles.' *International Journal of Control* 85 (7), pp. 880–97.

Shafer, G. (1990). 'The Unity and Diversity of Probability.' *Statistical Science 5* (4), pp. 435–562.

Shaki, S. and Fischer, M. (2008). 'Reading Space into Numbers: A Cross-Linguistic Comparison of the SNARC Effect.' *Cognition* 108, pp. 590–99.

Shaki, S. and Fischer, M. (2012). 'Multiple Spatial Mappings in Numerical Cognition.' *Journal of Experimental Psychology: Human Perception and Performance* 38 (3), pp. 804–9.

Spelke, E. (2011). 'Natural Number and Natural Geometry.' In: Brannon, E. and Dehaene, S. (eds), *Time and Number in the Brain: Searching for the Foundations of Mathematical Thought Attention & Performance* XXIV, pp. 287–317. Oxford: Oxford University Press.

Steiner, M. (1998). *The Applicability of Mathematics as a Philosophical Problem*. Cambridge, MA: Harvard University Press.

Stigler, S. (1986). *The History of Statistics: The Measurement of Uncertainty before 1900*. Cambridge, MA: Harvard University Press.

Syrett, K., Musolino, J. and Gelman, R. (2012). 'How Can Syntax Support Number Word Acquisition?' *Language Learning and Development* 8, pp. 146–76.

Tabak, J. (2004). *Probability and Statistics: The Science of Uncertainty*. New York: Facts on File.

The Economist (2017a). 'Crime and Despair in Baltimore: As America Gets Safer, Maryland's Biggest City Does Not.' *The Economist*, 29 June 2017. Online at https://www.economist.com/unitedstates/2017/06/29/crime-and-despair-in-baltimore.

The Economist (2017b). 'The Gender Pay Gap: Women Still Earn a Lot Less than Men, Despite Decades of Equal-Pay Laws. Why?' *The Economist*, 7 October 2017. Online at https://www.economist.com/international/2017/10/07/the-gender-pay-gap.

The Economist (2018). 'The Average American is Much Better Off Now than Four Decades Ago: Estimates of Income Growth Vary Greatly Depending on Methodology.' *The Economist*, 31 March 2018. Online at https://www.economist.com/finance-andeconomics/2018/03/31/the-average-american-is-much-better-offnow-than-four-decades-ago.

Vargas, J., López, J., Salas, C. et al. (2004). 'Encoding of Geometric and Featural Spatial Information by Goldfish (*Carassius auratus*).' *Journal of Comparative Psychology* 118 (2), pp. 206–16.

Wang, F. and Spelke, E. (2002). 'Human Spatial Representation: Insights from Animals.' *Trends in Cognitive Science* 6 (9), pp. 376–82.

Wassman, J. and Dasen, P. (1994). 'Yupno Number System and Counting.' *Journal of Cross-Cultural Psychology* 25 (1), pp. 78–94.

Wigner, E. P. (1960). 'The Unreasonable Effectiveness of Mathematics in the Natural Sciences.' *Communications on Pure and Applied Mathematics* 13 (1), pp. 1–14.

Wilson, M. (2000). 'The Unreasonable Uncooperativeness of Mathematics in the Natural Sciences.' *The Monist* 83 (2), pp. 296–314.

Winter, C., Kristiansen, G., Kersting, S. et al. (2012). 'Google Goes Cancer: Improving Outcome Prediction for Cancer Patients by Network-Based Ranking of Marker Genes.' *PLoS Computational Biology* 8 (5), e1002511.

Wynn, K. (1992). 'Addition and Subtraction by Human Infants.' *Nature* 358, pp. 749–50.

Xu, W. (2003). 'Numerosity Discrimination in Infants: Evidence for Two Systems of Representations.' *Cognition* 89, B15-B25.

INDEX